Praise for *Superagency*

"Artificial Intelligence is a set of ideas humanity has pondered since antiquity with a mix of trepidation and excitement. Now, in the early days of the 21st century we've digitized a huge fraction of human experience, and through rapid advances in computer science and basic computing infrastructure have finally started to realize some of the long promises of AI. *Superagency* is an important work, an encouragement to us all to think about how to harness the power of AI to amplify and enhance humanity. It asks us to imagine how to use this tool to give each of us a form of superagency with which we can create, connect, and invent more, and to nudge us to do what humans have always done: to become more essentially human through the use of our tools. The AI future that we build for ourselves will be entirely a function of our best collective imagination, our hopes and dreams for the problems we can solve with this technology, and the benefits we can create for ourselves and each other by making the most of AI's vast potential."

—Kevin Scott, CTO of Microsoft

"*Superagency* is a fascinating and insightful book, providing humanity with a bright vision for the age of AI. I disagree with some of its main arguments, but I nevertheless hope they are right. Read it and judge for yourself."

—Yuval Noah Harari, author of *Sapiens*

"Reid has been a great partner from the earliest days of OpenAI. Reid helped us navigate many challenges and opportunities as a board member, investor, and advisor. Superagency is an excellent vision to steer towards: where AI brings a renaissance of human agency and quality of life for everyone."

—Greg Brockman, President and Co-Founder of OpenAI

"*Superagency* is a refreshingly optimistic and welcome perspective on the future of AI. Rather than focusing on fear and disruption, Reid uses his 'techno-humanist compass' to show how AI can help us make better decisions, achieve our goals, and create a more equitable and sustainable world. This is a must-read for thriving in the age of AI."

— Arianna Huffington, Founder and CEO, Thrive Global

"At a time when artificial intelligence is generating justifiable fears and excitement, Reid Hoffman has emerged both as a trailblazing AI visionary and an invaluable thought leader on the nexus of technology and humanity. Reid has such deep knowledge about the human condition and a yearning for us to have a bolder moral imagination about how we can create a more caring, connected, and fulfilling society. For me, *Superagency* is a must read. Anyone who hopes this emerging technology, despite its potential challenges and shortcomings, can empower us all to create a more beloved community, would be wise to immerse themselves in Reid's wisdom."

—Senator Cory Booker

"Reid is one of the most pivotal thinkers and actors in the AI industry today. His consistent and grounded optimism about the potential here is a vital counterweight to a world that quickly forgets what benefits they can bring. In his brilliant new book *Superagency* he beautifully outlines a highly persuasive vision for what humanity can achieve at this critical juncture. A powerful and essential read for anyone who cares about AI, this is a rallying cry for a better future."

—Mustafa Suleyman, CEO of Microsoft AI
and Co-Founder Inflection AI and DeepMind

"Reid wields his own superintelligence to lay out a rationally optimistic case for the pivotal role AI can play as a partner in our lives. He also reinforces the idea that the sexiest thing in the world is to be really, really smart, and AI might just help us all be a little bit sexier. This is not the first time humanity has innovated, and it will not be the last."

—Ashton Kutcher, Actor, Investor, Father

"AI is the most important technology of our lifetimes. *Superagency* is one of the few books making the essential positive case that AI can and should massively benefit humanity. While we must navigate the risks, we need to have a future that we seek—and thus *Superagency*."

— Eric Schmidt, Former CEO of Google

"AI could compress a century of progress into a decade. *Superagency* explores this radical possibility, challenging us to consider how we might shape an AI-enabled world that amplifies human potential."

—Dario Amodei, CEO and Co-Founder, Anthropic

"*Superagency* is the right book at the right time—a valuable resource for readers who are interested in AI and ready to move beyond the barren strife of social media disputation. Reid Hoffman has unparallelled firsthand knowledge of the industry, combined with a commonsense mentality and a knack for lucid and enjoyable prose."

—Neal Stephenson

"Amid the Great Big AI Panic of 2024—with so many nervously calling for tight regulations, or moratoria, or outright renunciation—Reid Hoffman and a few others have beckoned us to recall what actually works. To seek a 'positive sum game,' where harms are caught and minimized and benefits multiply. For sure, it's never been easy, calling for agile mixtures of regulation, foresight, competition, and cooperation. But the potential lifesaving—and maybe planet-saving—benefits are too myriad to ignore."

—David Brin, Author of *The Postman* and *The Transparent Society*

"*Superagency* is the must-read guide to thinking about thinking now. Reid's sharp urgency and powerful insights come directly from the wave of human and machine fusion sparking all around us. From his unique position on the front lines of this revolution, Reid does more than teach us to contemplate our human/machine future. He tells us very specifically how to thrive."

—Joshua Cooper Ramo, Two-Time *New York Times* Bestselling Author of *The Age of the Unthinkable* and *The Seventh Sense*

"An inspiring picture of a future where AI doesn't just automate tasks, but amplifies human potential. This vision resonates deeply with my own passion for democratizing learning. *Superagency* is a call to action for educators, technologists, and policymakers to embrace AI's potential and work together to build a future where education truly empowers every individual to reach their full potential."

—Sal Khan, CEO Khan Academy

"AI has the potential to usher in an era of soaring progress, and *Superagency* shows us how. Reid Hoffman and Greg Beato offer a clear road map for harnessing the power of AI to transform work, education, and health care. A must read!

—Erik Brynjolfsson, Director of the Stanford Digital Economy Lab and Co-Founder of Workhelix

"Reid Hoffman is that rare Silicon Valley leader who seems suspicious of techno-utopianism but is nonetheless an unabashed evangelist for innovation as an indispensable propellant of human civilization. In *Superagency,* he makes a persuasive case for embracing our AI future—a world in which *Homo techne* uses this powerful new tool in the unending quest for a gentler and more meaningful existence."

— Dele Olojede, Winner of the Pulitzer Prize

"Reid has been at the center of this wave of AI innovation for nearly a decade, and I couldn't think of a kinder, more compassionate person to help shape our society. *Superagency* is a provocative, must-read for anyone interested in our shared AI future."

—Nancy Lublin, Founder of Dress for Success and Four-Time Entrepreneur

"In *Superagency,* Hoffman and Beato present a thoughtful road map for an AI-enabled future that enhances human agency. Their optimistic vision of AI as a tool for individual empowerment and democratic renewal is a welcome addition to the discourse. If you want a hopeful, nuanced take on how AI can help create a better future for all of us, this is it!"

—Ethan Mollick, Associate Professor of Management and Co-Director of the Generative AI Labs at Wharton

"AI is empowering individuals with newfound superpowers, yet far too much of the public discourse centers on doom and dystopia. In *Superagency,* Reid issues a powerful call to action for everyone to participate in creating a world where human ingenuity and machine intelligence work together to achieve extraordinary outcomes."

—Andrew Ng, Managing General Partner,
AI Fund and Founder, DeepLearning.AI

"Whether artificial intelligence is a threat or an opportunity depends on how we understand and use it. In his wonderful book *Superagency,* Reid Hoffman shows the way to make it part of a better, more prosperous future where we all benefit."

—Arthur C. Brooks, Harvard Professor
and #1 *New York Times* Bestselling Author

"Within the maelstrom of contemporary debate about what AI will do for us—and to us—Reid Hoffman offers deeply thoughtful and grounded reasons for optimism. This book invites both the skeptical and the bullish to refine a vision for a thriving future and how to build it, rather than simply waiting for what happens next."

—Jonathan Zittrain, Professor of Law, Computer
Science, and Public Policy, Harvard University;
Co-Founder, Berkman Klein Center for Internet & Society

"In *Superagency,* Hoffman and Beato remind us of a crucial truth: the AI we build today will define the world we see tomorrow. They show us that we have a choice—we can create AI that expands human potential and puts more power in people's hands, or we can build systems that limit our options. Drawing on lessons from past technologies, they make a compelling case for AI that amplifies individual agency rather than restricting it. This book shows that by making the right choices now, we can ensure that AI becomes a tool for personal empowerment and digital transformation."

—Kanjun Qiu, CEO, Imbue

"Much like Socrates's parable of the silent paintings, *Superagency* beckons us to break the stillness of fear and inertia, urging us to embrace AI as a new language of human agency, where iteration is the art of the future."

—Sean White, CEO, Inflection AI

"One of the smartest and most thoughtful entrepreneurs of our time analyzing the most interesting subject of our age. The world will be much better if this book is right."

—Rory Stewart, Yale Professor and Former UK Cabinet Minister

"AI has the potential to reshape societies across the globe in unprecedented ways, at unprecedented speeds. *Superagency* proposes a techno-humanist path forward to elevate humanity to new heights of individual fulfillment and shared prosperity."

—Qi Lu, Founder and CEO, MiraclePlus

"The AI revolution has arrived; we must now decide where it takes us. In *Superagency*, tech pioneer Reid Hoffman surveys a compelling, nuanced, and hopeful landscape—a world of innovation, opportunity, and ingenuity, transformed by powerful new technologies for the betterment of humanity."

—Laurene Powell Jobs

"Reid Hoffman is not only one of the most influential tech entrepreneurs and investors of our time—he's also one of our most prescient futurists. In this eye-opening book, he highlights the risks of not making advances in AI and reveals how it's poised to transform our lives. His insights will leave you more informed and excited about the revolution that's already underway."

—Adam Grant, #1 *New York Times* Bestselling Author of *Think Again* and *Hidden Potential,* and Host of the Podcast *Re:Thinking*

SUPERAGENCY

ALSO BY REID HOFFMAN

The Startup of You

The Alliance

Blitzscaling

Masters of Scale

Impromptu

SUPERAGENCY

What Could Possibly Go Right with Our AI Future

REID HOFFMAN
and
GREG BEATO

Authors Equity
1123 Broadway, Suite 1008
New York, New York 10010

Jacket design by Pete Garceau
Jacket images: Getty Images/iStockPhoto
Book design by Dix Digital Prepress and Design, Inc.

Some of the material and concepts found in this book have appeared in partial form in articles published by the authors.

Library of Congress Control Number: 2024947638
Print ISBN 9798893310108
Ebook ISBN 9798893310139

Printed in the United States of America
First Printing
www.authorsequity.com

CONTENTS

A NOTE ON THE TEXT

This book is a collaboration between my coauthor Greg Beato and me. We use "we" when representing our collective viewpoint. In instances specific to details from my own life, we revert to "I."

—*Reid*

SUPERAGENCY

INTRODUCTION

Throughout history, new technologies have regularly sparked visions of impending dehumanization and societal collapse. The printing press, the power loom, the telephone, the camera, and the automobile all faced significant skepticism and sometimes even violent opposition on their way to becoming mainstays of modern living.

Fifteenth-century doom-mongers argued that the printing press would dramatically destabilize society by enabling heresy and misinformation, and by undermining the authority of the clergy and scholars. The telephone was characterized as a device that could displace the intimacy of in-person visits and also make friends too transparent to one another.[1] In the early decades of the car's ascent, critics claimed it was destroying family life, with unmarried men choosing to save up for Model Ts instead of getting married and having kids, and married men resorting to divorce to escape the pressures of consumption that cars helped create.[2]

This same kind of doom and gloom was applied to society-wide automation in the 1950s, when increasingly sophisticated machines were dramatically impacting factories and office buildings alike, with everyone from bakers, meatcutters, autoworkers, and U.S. Census Bureau statisticians seeing their overall numbers dwindle. In 1961, *Time* magazine reported that labor experts believed that without intervention from business interests, unions, and the government, automation would continue to grow the "permanently unemployed."[3] By the mid-1960s, congressional subcommittees were regularly holding hearings regarding the mainframe computer's potential threat to privacy, free will, and the average citizen's capacity to make a life of their own choosing.

Today, U.S. unemployment rates are lower than they were in 1961. The average U.S. citizen lives in a world where PCs, the internet, and smartphones have ushered in a new age of individualism and self-determination rather than crushing authoritarian compliance or the end of humanity. But with the emergence and ongoing evolution of highly capable AIs, it's not just that familiar fears about technology persist; they're growing.

Even among AI developers, some believe that future instances of superintelligent AIs could represent an extinction-level threat to humanity. Others point out that, at the very least, humans acting with malicious intent will be able to use AIs to create catastrophic damage well before the machines themselves wage unilateral war against humanity. Additional concerns include massive job displacement, total human obsolescence, and a world where a tiny cabal of techno-elites capture whatever benefits, if any, AI enables.

The doomsday warnings are different this time, these observers insist, because the technology itself is different this time. AI can already *simulate* core aspects of human intelligence. Many researchers believe it will soon attain the capacity to act with complete and extremely capable autonomy, in ways that aren't aligned with human values or intentions.

Robots and other kinds of highly intelligent systems have long existed in sci-fi novels, comic books, and movies as our dark doppelgangers and adversaries. So as today's state-of-the-art AIs hold forth like benevolent but coolly rational grad students, it's only natural to see foreshadowing of HAL from *2001: A Space Odyssey*, or the Borg from *Star Trek*, or, in a less self-aware and more overtly menacing form, *The Terminator*'s relentless killer robot. These narratives have shaped our worst visions of the future for a long, long time.

But are they the right narratives? The future is notoriously hard to foresee accurately—for pessimists and optimists alike. We didn't get the permanent mass unemployment that labor experts in the early 1960s anticipated; nor did we get *The Jetsons* and its flying cars—at least not yet.

As hard as it may be to accurately predict the future, it's even harder to stop it. The world keeps changing. Simply trying to stop history by entrenching the status quo—through prohibitions, pauses, and other efforts to micro-manage who gets to do what—is not going to help us humans meet either the challenges or the opportunities that AI presents.

That's because as much as collaboration defines us, competition does too. We form groups of all kinds, at all levels, to amplify our efforts, often deploying our collective power against other teams, other companies, other countries. Even within our own groups of like-minded allies, competition emerges, because of variations in values and goals. And each group and subgroup is generally adept at rationalizing self-interest in the name of the greater good.

Coordinating at a group level to ban, constrain, or even just contain a new technology is hard. Doing so at a state or national level is even harder. Coordinating globally is like herding cats—if cats were armed, tribal, and had different languages, different gods, and dreams for the future that went beyond their next meal.

Meanwhile, the more powerful the technology, the harder the coordination problem, and that means you'll never get the future you want simply by prohibiting the future you *don't* want. Refusing to actively shape the future never works, and that's especially true now that the other side of the world is only just a few clicks away. Other actors have other futures in mind.

What should we do? Fundamentally, the surest way to prevent a bad future is to steer toward a better one that, by its existence, makes significantly worse outcomes harder to achieve.

At this point we know from thousands of years of experience that if a technology can be created, humans will create it. As I've written elsewhere, including in my previous book, *Impromptu*, we're *Homo techne* at least as much as we're *Homo sapiens*. We continuously create new tools to amplify our capabilities and shape the world to our liking. In turn, these tools end up shaping us as well. What this suggests is that humanism and technology, so often presented as oppositional forces, are in fact integrative ones. Every new technology we've invented—from language, to books, to the mobile phone—has defined, redefined, deepened, and expanded what it means to be human.

We're the initiators of this process, but we can't fully control it. Once set in motion, new technologies exert a gravity of their own: a world where steam power exists works differently than the world that preceded it. This is precisely why prohibition or constraint alone is never enough: they offer stasis and resistance at the very moment we should be pushing forward in pursuit of the brightest possible future.

Some might describe this as technological determinism, but we think of it as navigating with a kind of techno-humanist compass. A compass helps us to choose a course of action, but unlike a blueprint or some immutable manifesto, it's dynamic rather than determinative. It helps us orient, reorient, and *find* our way.

It's also crucial that this compass be explicitly humanist, because ultimately every major technological innovation impacts human agency—our ability to make choices and exert influence on our lives.

A techno-humanist compass actively aims to point us toward paths in which the technologies we create broadly augment and amplify individual and collective agency.

With AI, this orientation is especially important. Because what happens to human agency when these systems and devices, often described as agents themselves, do become capable of replacing us entirely? Shouldn't we slow down that eventuality as much as possible? A techno-humanist perspective sees it the other way around: our sense of urgency needs to match the current speed of change. We can only succeed in prioritizing human agency by actively participating in how these technologies are defined and developed.

First and foremost, that means pursuing a future where billions of people around the world get equitable, hands-on access to experiment with these technologies themselves, in ways of their own choosing. It also means pursuing a future where the growing capabilities of AI help us reduce the threats of nuclear war, climate change, pandemics, resource depletion, and more.

In addition, it means pursuing this future even though we know we won't be able to predict or control every development or consequence that awaits us. No one can presume to know the exact final destination of the journey we're on or the specific contours of the terrain that exists there. The future isn't something that experts and regulators can meticulously design—it's something that society explores and discovers collectively. That's why it makes the most sense to learn as we go and to use our techno-humanist compass to course-correct along the way. In a nutshell, that's "iterative deployment," the term that OpenAI, ChatGPT's developer, uses to describe its own method in bringing its products into the world. It's a concept my coauthor, Greg Beato, and I will explore and emphasize throughout this book.

As a longtime founder and investor in technology companies, my perspective is inevitably shaped by the technology-driven progress and positive outcomes I've participated in over the course of my career. I was a founding board member at PayPal and part of its executive team when eBay purchased it in 2002. I cofounded LinkedIn and have sat on Microsoft's board since 2017, following its purchase of LinkedIn.

I was also one of the first philanthropic supporters of OpenAI when it launched as a nonprofit research lab in 2015. I led the first round of investment in 2019 when OpenAI established a for-profit limited partnership in order to support its ongoing development efforts. I served on its board from 2019 to early 2023. Along with Mustafa Suleyman, who cofounded DeepMind, I cofounded a public benefit corporation called Inflection AI in 2022 that has developed its own conversational agent, Pi. In my role at the venture capital firm Greylock, I've invested in other AI companies. On my podcast *Possible*, I regularly talk with a wide range of innovators about the impacts AI will have on their fields—with a techno-humanist compass guiding our conversations. I also provide philanthropic support to Stanford University's Institute for Human-Centered Artificial Intelligence (HAI) and to the Alan Turing Institute, the United Kingdom's national institute for data science and artificial intelligence.

I recognize that some might say such qualifications actually disqualify my perspective on AI. That my optimism is merely hype. That my idealism about how we might use AI to create broad new benefits for society is just an effort to generate economic return for myself. That my roles as founder, investor, advisor, and philanthropic supporter of many AI-focused companies and institutions create an ongoing incentive for me to overpromote the upsides and downplay the dangers and downsides.

I argue that the opposite is true: I'm deeply involved in this technology and I want to see it succeed exactly because I believe it can have profoundly positive impacts on humanity. My engagement in this domain has meant that I've seen firsthand the progress being made. That has strengthened my commitment, and thus I've continued to invest in and support a widening range of companies and organizations. I stay alert to potential dangers and downsides, and am ready to adapt, if necessary, precisely because I want this technology to succeed in ways that broadly benefit society.

One reason iterative deployment makes so much sense in the case of pioneering technologies like AI is that it favors flexibility over some grand master plan. It makes it easier to change pace, direction, and even strategy when new evidence signals the need for that.

Meanwhile, here we are presenting our argument to you in a book.

Roughly 2,400 years ago, Socrates critiqued the written word for its lack of dynamism in Plato's *Phaedrus,* and for the way it made knowledge accessible to anyone:

> You know, Phaedrus, writing shares a strange feature with painting. The offsprings of painting stand there as if they are alive, but if anyone asks them anything, they remain most solemnly silent. The same is true of written words. You'd think they were speaking as if they had some understanding, but if you question anything that has been said because you want to learn more, it continues to signify just that very same thing forever. When it has once been written down, every discourse rolls about everywhere, reaching indiscriminately those with understanding no less than those who have no business with it, and it doesn't know to whom it should speak and to whom it should not.[4]

For Socrates, apparently, fixing his thoughts into written text represented a loss of agency. Had he turned his teachings into books himself, or rather scrolls, the reigning technology of his day, he would not have been able to control who read them. He would not have always been on hand to provide updates on his thinking, elaborate on nuances in the text, or correct misreadings. Consequently, face-to-face dialogic inquiry was his preferred technology for transmitting ideas.

But clearly generations of authors and readers thought differently. Why? Because ultimately written works increased the agency of authors and readers, enabling the latter to engage with, learn from, modify, expand upon, and, yes, perhaps even misinterpret or appropriate ideas from authors with which they might never have otherwise crossed paths.

As printing technologies improved, books evolved into a transformative global resource. Rolling about everywhere, indiscriminately reaching everyone, they functioned as early mobility machines, decoupling human cognition from human brains, democratizing knowledge, accelerating human progress, and providing a way for individuals and whole societies to benefit from the most profound and impactful human insights and innovations across time and space.

Of course, there are myriad other ways to share information now, and we'll be using many of them to convey the ideas in *Superagency* too. Along with the usual podcasts and social media, we'll be experimenting with AI-generated video, audio, and music to augment and amplify the key themes we're exploring here. To see how, check our website Superagency.ai.

But we're starting with a book—in part as homage to the essential truth that technologies that often seem decidedly flawed and even dehumanizing at first usually end up being exactly the opposite.

CHAPTER 1

HUMANITY HAS ENTERED THE CHAT

As 2022 drew to a close, people around the world continued to navigate the complex landscape of postpandemic recovery. Polls showed that inflation had replaced Covid-19 as the top global concern.[1] Food prices remained at record highs. But the return to normalcy was also in full swing. Job growth and average hourly wage increases for the month of November would greatly exceed estimates.[2] Tickets to Taylor Swift's concert tour were in such high demand that Ticketmaster's systems buckled under the weight of 14 million simultaneous customers.

Still, the tech industry itself was having some trouble shaking off the cumulative effects of advertising slowdowns, shifting investor sentiments, and evolving user engagement patterns. On November 9, Meta, the company formerly known as Facebook, laid off 11,000 employees, its largest-ever workforce reduction. Two days later, FTX, the Bahamas-based cryptocurrency exchange, declared bankruptcy with allegations of massive fraud and misuse of customer funds quickly following. The news rocked the entire cryptocurrency ecosystem, wiping out billions in market value. On November 14, the bad news continued, with the report that, like Meta, Amazon had begun to conduct unprecedented layoffs, with at least 10,000 employees to be let go in the coming weeks.

In part, these adverse outcomes were simply a correction to the surges in tech industry hiring, revenue, and market caps that pandemic stimulus and pent-up consumer demand had inspired. But they also provided a clear rebuttal to the ongoing narrative regarding Big Tech's alleged power to exercise complete control over markets and manipulate consumer behavior at will.

Throughout the 2010s, this narrative had effectively become gospel. Through addictive design practices and an arsenal of attention-hijacking techniques, it asserts, companies like Alphabet (aka Google) and Meta have all but perfected the dark art of maximizing engagement and trapping millions of users in digital hells of doomscrolling, rage-tweeting, shit-posting, rabbit-holing, thirst-trapping, hate-reading, group-shaming, and self-retweeting.

But none of Meta's allegedly irresistible tactics had managed to lure many people to the metaverse, its multibillion-dollar effort to create a vast virtual world where users could work, play, and socialize in immersive 3-D environments. And while Silicon Valley venture capitalists had been pouring billions into blockchain startups and other cryptocurrency projects, the developers of these technologies had not yet cracked the code for making decentralized financial as indispensable to mainstream users as the web, search, email, mobile computing, text messaging, and social media had quickly become in earlier waves of digital innovation.

Meanwhile, on the final day of November, after a month that had played out as one long swipe-left for the tech industry, OpenAI, a San Francisco–based research lab with a staff of around 375 employees, unveiled its latest product with no advance notice and zero hype. "Try talking with ChatGPT, our new AI system which is optimized for dialogue," the lab's official X.com account tweeted[3] at 10:02 a.m. "Your feedback will help us improve it."

OpenAI's cofounder and CEO, Sam Altman, was similarly circumspect in his first tweet[4] about ChatGPT: "language interfaces are going to be a big deal, i think. talk to the computer (voice or text) and get what you want, for increasingly complex definitions of 'want'! this is an early demo of what's possible (still a lot of limitations—it's very much a research release)."

As its name suggested, ChatGPT was a chatbot—not exactly many people's first pick to be the next *Pokémon Go*, or even the next Juicero. For most of their history on the web, where they were typically put to use in customer service contexts, chatbots functioned so poorly that the average person looking for guidance on how to return an online purchase still preferred to wait on hold for twenty minutes to speak to someone who sounded like they were answering from a call center located in the bottom of a swimming pool on Mars.

But ChatGPT was different. Very different. And that difference was immediately apparent. Impressively knowledgeable, stunningly versatile, and convincingly human, ChatGPT could provide an easy-to-understand explanation of quantum mechanics. It could compose sonnets about the Consumer Price Index, if that's what you wanted. It could help you debug—or possibly bug—your Python code. ChatGPT didn't always get things right, but even its mistakes, commonly described as "hallucinations," induced wonder and intrigue.

With zero marketing dollars behind it, ChatGPT attracted its first one million users in five days. Somehow, it seemed like two million of those one million people were journalists, reporting on their experiences—and that's when interest in ChatGPT really started to skyrocket. In just two months, it attracted 100 million users[5] and generated so much excitement, aspiration, and FOMO in the tech industry that it should have gotten a commendation from the Federal Trade Commission's Bureau of Competition.

Alphabet CEO Sundar Pichai sent a code-red alert to every Googler announcing that AI was now Job 1 for the entire company. Microsoft, which had invested in OpenAI three years earlier, started mentioning copilots more often than an airline training manual. Mark Zuckerberg announced that Meta had created a new top-level generative AI product group to "turbocharge" the company's efforts in the field.[6] Newcomers and upstarts like Anthropic, Midjourney, Hugging Face, and Replika pushed AI forward in ways that left the giants trying to keep up. Even research papers with titles along the lines of "Quantum Entanglement of Neural Networks in Multidimensional Latent Spaces" could go low-key viral on X.com.

In the face of these new conditions, sentiments took a 180-degree turn. For years prior to ChatGPT's release, Big Tech's biggest critics had been insisting that antitrust actions were necessary to inject new competitive energy into this once-dynamic U.S. technology sector. Now those same critics started saying that innovation was happening too quickly, that it was out of control. In March 2023, a nonprofit organization called the Future of Life Institute published an open letter that urged "all AI labs to immediately pause for at least 6 months the training of AI systems more powerful than GPT-4."[7] More than 33,000 people, including many AI-industry leaders and technologists, signed it. Their mood was dire, their urgency palpable. The Senate Judiciary Committee took note and conducted multiple hearings on AI oversight throughout the year.

Six months came and went, then another six months. Developers kept working, and the innovations continued. OpenAI continued to release updates to GPT-4, the foundation model underlying ChatGPT that was achieving state-of-the-art performance on complex tasks and problem-solving, and gaining the capacity to analyze images and provide feedback on them. Anthropic's Claude 2 achieved new levels of factual accuracy and expanded its context length, meaning it could process and keep track of context in input texts of up to about 75,000 words. If you wanted to summarize the complete and unabridged version of H. G. Wells's *The War of the Worlds* in twenty bullet points, Claude could do that.

Still, many of the ongoing challenges remained. Conversational agents like ChatGPT and Claude are built on top of large language models, or LLMs, a specific kind of machine learning construct designed for language-processing tasks. LLMs like GPT-4 process and generate language using what's known as neural network architecture, in which multiple layers of nodes perform a complex cascade of interconnected computations. Each node in a layer takes input from the previous layer, applies mathematical operations, and passes the result to the next layer. Parameters, as they're called, are also pivotal in this process, as they determine the strength of connections between nodes.

In a process known as pretraining, LLMs learn associations and correlations between *tokens*—words or fragments of words—by scanning a vast amount of text. In an LLM, each parameter functions something like a tuning knob, and in today's largest models, there are hundreds of billions of them. Through an iterative process of adjusting these parameters across all nodes in the network, the model reinforces or reduces connections between the tokens in its training data and begins to recognize and replicate complex patterns in language.

What this means, in part, is that LLMs never know a fact or understand a concept in the way that we do. Instead, every time you prompt an LLM with a question, or ask it to take some action, you are simply asking it to make a prediction about what tokens are most likely to follow the tokens that comprise your prompt in a contextually relevant way. And they don't always make correct or appropriate predictions.

By expanding pretraining datasets, fine-tuning model performance on more task-specific datasets, along with other measures, developers try to make their models more accurate and less prone toward undesirable outputs. As models grow more capable, they begin to display a sophisticated kind of simulated "awareness" of the world, such as recognizing that when a person says "I'm so hungry I could eat a horse" they're using hyperbolic language for expressive impact rather than asking for horse recipes.

But even when it seems like models possess humanlike commonsense reasoning, they don't. Instead they're making statistically probable predictions regarding patterns of language. This means they sometimes make mistakes. They can behave unpredictably. When models generate false information or misleading outcomes that do not accurately reflect the facts, patterns, or associations grounded in their training data, they are said to be "hallucinating"—that is, they're "seeing" something that isn't actually there.*

That means a model might provide an incorrect answer to a question that has a correct answer. It might fabricate entirely novel "facts," such as names, dates, or events, that have no basis in reality. It can provide information that may be accurate but has no contextual relevance to a given user prompt. Finally, it can generate outputs that are logically inconsistent or incoherent.

In addition, a model's dependence on data and quantification may give it the appearance of objectivity or neutrality, but it's not objective or neutral. Instead it's created by human developers and institutions

* Some AI researchers assert that "confabulation," which means creating false memories or narratives to fill gaps in knowledge, is a more appropriate description than "hallucination," which suggests perceiving something that isn't there. Both are instances of anthropomorphizing AIs, attributing human characteristics to nonhuman entities. We use "hallucination" in this book because it is the most commonly used term for the phenomenon.

making choices about which data to collect, how to process that data, what purpose or specific function a model is being optimized for, how best to align that function with human values and intents, and so on.

If a model's training data contains sexist or racist sentiments—which can happen when massive quantities of text are scraped from the internet and subjected to little or no additional filtering, vetting, or refinement—then the model might produce sexist or racist outputs. If a developer creating an AI for medical diagnosis doesn't fully grasp the complexities of certain medical conditions, that could lead to models that perform poorly for underrepresented patient groups or rare diseases.

Another issue involves the often opaque ways large language models operate, a characteristic known as the "black box" phenomenon. This occurs when complex neural networks processing hundreds of billions of text samples in extremely granular fashion identify patterns that human overseers have trouble discerning—making it hard or even impossible to explain a model's outputs or trace its decision-making process.

While developers apply various techniques to mitigate these issues, the fundamental limitation that underlies them all remains the same. As of yet, LLMs have no real capacity for commonsense reasoning, no lived experience, and no grounded model of the world. They're always just predicting the next token in a sequence, based on patterns they've learned from their training data.

So even when built on top of a state-of-the-art language model, ChatGPT and its peers continue to hallucinate. They can still tie themselves in knots trying to solve relatively simple brain-teasers. They sometimes generate biased outputs in some instances and contextually irrelevant ones in others.

And this won't ever completely change, skeptics assert. Even as developers build bigger supercomputers on which to train their models with trillions of parameters, and expand training datasets to include multimodal inputs like images, videos, and structured data, performance gains have started to slow. Errors persist. Critics say they'll never achieve AI's holy grail—*artificial general intelligence,* or AGI, in which models become capable of applying knowledge from one domain or context to entirely different situations, adapting to new challenges with humanlike flexibility, reasoning abstractly across diverse fields, and generating original ideas and solutions—all without being explicitly programmed for each task.

The Beginning Is Near

Throughout 2023, thanks to LLMs, AI was the biggest story in tech. Many observers believed these new models were on the verge of changing everything, in a good way. Many others believed they were on the verge of changing everything, in cataclysmically bad ways. Still others believed they were on the verge of keeping everything the same, only more so, in terms of concentrating power, profits, and the future of the world in the hands of a few Big Tech players.

By the summer of 2024, however, an ironic shift had occurred. Whereas generative AI critics had once demanded a six-month pause in the development of systems more powerful than GPT-4, for fear of potential catastrophic risks, there were now questions about why it was taking so long to deliver the next generation of groundbreaking models. And along with those questions there were persistent and increasing doubts regarding how much more capable LLMs might ultimately get. Thus, what had once been portrayed as Public Enemy No. 1 was now being deemed a dud. Phrases like "AI hype," "AI bubble," and "troughs of disillusionment" began to seep into media headlines.

But I'd already been through similarly dizzying changes in expectations—in the opposite direction. In 2015, when I first became involved with OpenAI, the idea that AI could eventually achieve, or credibly simulate, humanlike understanding and reasoning remained on the fringes of conventional wisdom. Even in Silicon Valley, the prospect was considered an extreme long shot. That was one of the reasons OpenAI had established itself as a nonprofit. Venture capital firms seeking a return on investment within the traditional VC time frame of five to ten years would have been unlikely to commit to such a speculative and long-term endeavor.

And if there were big challenges to overcome in 2024, well, there had been big challenges in 2015 too. There were big challenges in 2018 and 2020. Somehow OpenAI and the rest of the AI development community had always managed to find new techniques and breakthroughs that enabled the next wave of improvement.

Granted, it's easy to be an optimist if your time horizons are long. And mine are. In fact, I believe that we're still in the very early stages of this new phase of human discovery and growth. The supercomputers are going to get even more super. Developers will continue to write more efficient algorithms. To overcome some of the limitations that characterize LLMs, they'll come up with new architectures and techniques, and incorporate different approaches like multimodal learning and neurosymbolic AI—systems that integrate neural networks with symbolic reasoning based on explicit, human-defined rules and logic.

What all that means is that we're also still in the early stages of the existential reckoning that these systems will provoke, as we try to process what it really means to introduce new, and not entirely predictable, forms of intelligence into our world.

A machine that can think like a human—strategically, abstractly, and even creatively, at the speed and scale of a computer—will obviously be revolutionary. What if every child on the planet suddenly has access to an AI tutor that is as smart as Leonardo da Vinci and as empathetic as Big Bird? What if billions of people around the world suddenly have a highly knowledgeable and reliable health care advisor in their pocket at all times?

Of course, not everyone focuses on the potential upsides of AI—especially in the U.S., where people consistently express high levels of concern about the technology. According to a 2022 survey conducted by Ipsos, a global analytics firm, "only 35% of sampled Americans (among the lowest of surveyed countries) agreed that products and services using AI had more benefits than drawbacks."[8] A similar survey from the Pew Research Center found that only 15 percent of U.S. adults said they were "more excited than concerned about the increasing use of AI in daily life."[9] In another study, from Monmouth University, 56 percent said that "artificially intelligent machines would hurt humans' overall quality of life."[10]

Such concerns are entirely understandable. We're experiencing a world-changing moment that's creating significant uncertainty. Exactly how good will these systems get, and how fast? What kinds of jobs will be left for people as AI continues to improve? What happens to trust and public discourse, already under siege, as AI technologies make it cheaper and easier to produce convincing simulations of reality at scale? What happens to individual privacy and autonomy in a highly instrumented world, where billions of systems, devices, and robots can match or exceed human performance, and can take actions that may infringe on our own choices and desires? Can we continue to maintain control of our lives, and successfully plot our own destinies?

Most Concerns About AI Are Concerns About Human Agency

It's this last question that informs this book's primary focus: Can we continue to maintain control of our lives, and successfully plot our own destinies? Ultimately, questions about job displacement are questions about individual human agency: Will I have the economic means to support myself, and opportunities to engage in pursuits I find meaningful? Questions about disinformation and misinformation are questions about individual human agency: How do I know whom and what to trust as I make decisions that impact my life? Questions about privacy are questions about individual human agency: How do I maintain the integrity of my own identity and how I'm known in the world, and preserve an authentic sense of self?

Human agency is a fundamental concept in philosophy, sociology, and psychology. It holds that you, as an individual, have the capacity to make your own choices, act independently, and thus exert influence over your life. While you may also believe that external circumstances and conditions play a significant role in the outcomes you experience, it's your sense of agency that compels you to form intentions, set goals, and take actions to achieve those outcomes. A sense of agency, then, can endow your life with purpose and meaning.

As AI systems evolve, their capacity for self-directed learning, problem-solving, and executing complex series of tasks without constant human oversight is increasing. While a self-driving car isn't conscious of its agency in the same way that humans are, it does have the capacity to make decisions autonomously, take actions, and pursue goals within its own operational domains.

In time, this means that an increasing variety of systems, devices, and machines will encroach on areas traditionally governed by human agency—in ways that humans may often find objectionable. And even in instances where we welcome such cognitive offloading, other issues arise: What if, through overreliance on machine agency and capabilities, our own skills and agency atrophy over time? What if the systems that are supposedly working on our behalf—and delivering outcomes we approve of—end up shaping our behaviors and choices in ways that we haven't explicitly consented to?

Well, that's actually the story of humanity to date. As *Homo techne,* we are defined by our capacity and commitment to creating new ways of being in the world through our tool-making. This is what distinguishes us from every other living being on the planet, even other rudimentary tool-makers. While generation after generation of chimpanzees might use stone tools to crack nuts, we consistently create *new* technologies, which we then use to exercise our agency in novel and increasingly productive ways. From stone tools to smartphones, this virtuous cycle of innovation expands and compounds what it means to be human over time.

Now we have an opportunity to develop new supertools. Ones that can increase our agency so much that we're now in the midst of something akin to the Industrial Revolution. The implications of that assertion are far-reaching, but here's one way to think about it: Working in tandem, intelligence and energy drive human agency, and thus human progress. Intelligence gives us the capacity to weigh options, and to envision and plan for different potential scenarios. Energy enables us to then take action on whatever we aspire to achieve. The more intelligence and energy we can leverage on our behalf, the greater our capacity to make things happen, individually and collectively.

Spoken language, controlled use of fire, the wheel, and written language were all pivotal technologies our ancestors used to augment and amplify human intelligence and energy. As we continue to add new innovations to the mix, this process continues. In the early twentieth century, automobiles powered by internal-combustion engines enabled millions of people to routinely achieve what would have been considered superhuman feats just a few decades earlier. In more recent decades, personal computers, the internet, and smartphones were similarly transformative in how they augmented and amplified human intelligence.

AI will enable our next great leap forward. In contrast to innovations like books or how-to videos on YouTube, AI isn't just a way to manufacture and distribute knowledge, as valuable as that is. Because an AI has the capacity to be agentic itself, setting goals and taking actions on its own to achieve them, you can leverage AI in two distinct ways. In some instances, you might want to work closely with an AI—such as when you're learning a new language or practicing mindfulness skills. In others, such as optimizing your home's energy consumption based on real-time energy prices and weather forecasts, you might prefer to let an AI handle that by itself.

Either way, the AI is increasing your agency, because it's helping you take actions designed to lead to outcomes you desire. And either way, something new and transformative is happening. For the first time ever, synthetic intelligence, not just knowledge, is becoming as flexibly deployable as synthetic energy has been since the rise of steam power in the 1700s. Intelligence itself is now a tool—a scalable, highly configurable, self-compounding engine for progress.

If we harness it correctly, we can achieve a new state of superagency. That's what happens when a critical mass of individuals, personally empowered by AI, begin to operate at levels that compound through society. In other words, it's not just that some people are becoming more informed and better equipped thanks to AI. Everyone is, even those who rarely or never use AI directly. Because suddenly your doctor can diagnose your vague and seemingly unrelated complaints with AI precision. Your auto mechanic knows exactly what that weird thump coming from your trunk means when you accelerate from a traffic light on a hot day. Even ATMs, parking meters, and vending machines are multilingual geniuses who understand your needs instantly and adjust to your preferences. That's the world of superagency.

Full Steam Ahead

Prior to the Industrial Age, productivity across society was greatly constrained because energy was a scarce commodity. Tilling fields, digging tunnels, operating pulleys—everything required extensive human or animal labor, limiting scale, efficiency, and, more broadly, human flourishing, as man and horse alike effectively functioned as beasts of burden. Windmills and water-powered mills enabled more efficient production of grain and textiles in some regions, nature permitting. If you were born near a river, life might be slightly more prosperous and offer you greater opportunities for self-expression. If not, well, hopefully you found some joy in plowing.

In the mid-1700s, though, things began to change with the rise of steam power.

In retrospect, we're often quick to view the Industrial Revolution through the worst effects that stemmed from it. Cities blackened by the coal-burning that steam power required. Grueling work in dangerous factories where workers had few if any rights. Child labor. More regimented and less communal living patterns. Overcrowding amid slum conditions and urban anomie at the same time.

You could even say it was a violently dehumanizing era. Factory clocks and streetlamps displaced nature's intrinsic rhythms. People were often expected to function more like machines. Consumer transactions in an increasingly commercialized world crowded out opportunities for mutual reciprocity and personal trust. Growth and prosperity on a global scale was uneven, with new demands for natural resources intensifying already-exploitative colonial relations and creating entirely new inequities.

But these weren't the only truths about steam power and the Industrial Revolution. Overall, across centuries, synthetic energy and industrialization have been radically humanizing forces. Steam-powered mechanization created opportunities to pursue scale at unprecedented levels. The pursuit of scale required new levels of cooperation and collaboration. Once achieved, scale produced new levels of abundance and variety.

A more technologically driven society also required a more educated populace. More people with a greater variety of skills and expertise led to additional innovations, which in turn inspired new possibilities and opportunities. New efficiencies in the production of food, goods of all kinds, and information ultimately helped create the social progress that often comes with prosperity and abundance: fairer laws, greater economic and cultural mobility, increasingly robust social welfare systems, and a growing emphasis on individual rights and freedoms. There was a time when almost all human effort was focused on growing food and slaughtering livestock. In augmenting literal manpower with synthetic manpower, the steam engine expanded the possibilities of what it meant to be human, in ways that were exponential in scope. It provided a path forward toward a more enriching, diverse, dignified, and humane existence.

Now all that can play out again, with AI. We can make an exponential leap forward, with synthetic intelligence having the same impact in the twenty-first century and beyond that synthetic energy began to have in the 1700s. Just as the Industrial Revolution created new opportunities for collaboration and new capacities for innovation, creativity, and productivity, this cognitive revolution will as well. As this transformation occurs, education and skill development will become even more critical, leading to a more knowledgeable and capable populace. And ultimately synthetic intelligence can expand human potential and human agency in the same way that synthetic energy has. It's a path toward a more fulfilling and humane existence.

AI That Works for Us and with Us

So what's the best way to pursue this positive vision of abundant intelligence centered on individual human agency? When OpenAI launched in 2015, it didn't have a specific product or service in mind, much less a business model or a go-to-market strategy. Instead, as two of OpenAI's cofounders, Greg Brockman and Ilya Sutskever, said in an essay published at launch, OpenAI's broad mandate was to "advance digital intelligence in the way that is most likely to benefit humanity as a whole."[11]

To achieve this ambitious but fairly abstract goal, Brockman and Sutskever suggested, OpenAI would follow a directive that would prove to be transformative: "We believe AI should be an extension of individual human wills and, in the spirit of liberty, as broadly and evenly distributed as possible."

Why was this perspective so pivotal? Even in 2015, some iterations of AI were in fact already broadly distributed. Most internet users were already using AI in the form of recommendation systems, news feed curation, and predictive text and auto-complete services, to name just a few applications. There were other scenarios too, often controversial. In 2016, ProPublica reported on an algorithm, in use throughout the

U.S. criminal justice system, that was "particularly likely to falsely flag black defendants as future criminals, wrongly labeling them this way at almost twice the rate as white defendants."[12]

That same year, a coalition of seventeen civil rights organizations issued a statement condemning predictive policing tools, which use algorithms and historic crime report data to forecast where and when crimes are likely to occur.[13] A few years later, in 2019, San Francisco became the first city to ban its police department and other city agencies from using facial recognition systems.[14]

All of these use cases shared a common thread: none of the people that AI was impacting in these scenarios had explicitly chosen to use it.

Even with recommendation services and news curation, you often aren't making an explicit choice to use AI. Instead you're simply following the path that developers have laid out for you. *Amazon says that I might be interested in this one thing because I just bought this other thing? Maybe I'll check it out.*

The new AI models that OpenAI began to release in 2020, starting with an early LLM called GPT-2, were profoundly different. They aren't embedded in other systems or activated without your knowledge. They're something you must affirmatively choose to use. Once you do, you can use them in open-ended, self-determining ways.

With the release of ChatGPT in November 2022, this approach achieved a new level of versatile utility. Sure, you could ask it to write an essay for you. But you could also ask it to critique an essay that you wrote. You could have it compose questions from a company where you had an upcoming interview. You could ask it to suggest a challenging but quiet workout routine you could do in your apartment without waking your roommate.

Finally, there was a highly accessible, easy-to-use AI tool that explicitly worked *with you* and *for you*, rather than *on you*. This marked a critical shift in AI development and human empowerment. It's critical because it puts individual users at the heart of the experience. And, just as important, it gives them opportunities to have experiences that they've sought or designed. Instead of developing this new technology behind closed doors until a small cadre of experts had decided that it was performing in sufficiently effective and perfectly safe ways, OpenAI invited the public to participate in the development process. It described this approach as iterative deployment.

Iterative deployment has sound technological, regulatory, and sociological underpinnings. In classic Silicon Valley style, it relies on user experience and explicit user feedback to inform ongoing development efforts; that's the technological part. It is attentive to regulatory needs, in that it introduces change incrementally, in part to limit the impact of potential adverse effects. It's sociologically minded in that it gives individuals and society alike time to adapt to these changes.

In essence, iterative deployment recognizes that Rome wasn't built in a day—and Roman citizens weren't either. It takes time to understand the contours of a new world and how to function in it. Trust starts with exposure and evolves with use. Once you learn what something is and how it functions, you begin to trust it. Trust equals consistency over time.

In the context of AI, we first must develop trust in the technologies themselves—no easy feat when the technologies are somewhat unpredictable and capable of error. If we have access to the tools, though, we can form opinions for ourselves: *Is it reliable enough to do the things I want to do? Does it offer genuine value or just novelty? Does it empower me or make me overly dependent?*

Trust in the technologies is just the start, though. We must also cultivate trust in the developers of the technologies, the regulators of the technologies, and, perhaps most of all, the other users of the technologies. After all, why should you trust other people to use AI in primarily positive ways? Why should you trust your government—or the governments of other countries—to do the same? Why should you trust AI developers to do their work competently and responsibly?

There's no short answer to these questions. The answer is an ongoing conversation, often contentious, just as it's always been when humanity creates new, transformative technologies. This particular conversation has been going on for years now, but mostly playing out among computer scientists, higher education researchers, commercial developers, investors, ethicists, technology journalists, policymakers, legal experts, and sci-fi writers. So far, at least four key constituencies have been informing the discourse: the Doomers, the Gloomers, the Zoomers, and the Bloomers.

While these are clearly broad-brush labels, we're not trying to reduce the complexity of the perspectives that underlie them. Instead, we simply aim to summarize the basic positions that have animated the debates around AI. All these schools of thought contribute unique and valuable insights, challenge assumptions, and push the boundaries of our collective understanding.

Doomers believe we're on a path to a future where, in worst-case scenarios, superintelligent, completely autonomous AIs that are no longer well aligned with human values may decide to destroy us altogether, except perhaps for a small contingent of tech bros whom they'll keep around to do the vacuuming as revenge on behalf of Roombas.

Gloomers are both highly critical of AI and highly critical of Doomers. In their estimation, the Doomer outlook serves dual purposes. First, it exists as a tacit endorsement of AI—"It's so powerful it just might destroy us!" Second, its long-term and abstract nature misdirects attention toward the future, when our real priority should be more near-term AI risks and harms such as potential job losses, disinformation on a massive scale, amplification of existing systemic biases, and the undermining of individual agency. In general, Gloomers favor a prohibitive, top-down approach, where development and deployment should be closely monitored and controlled by official regulation and oversight bodies.

In contrast, ***Zoomers*** argue that the productivity gains and innovation AI will create will far exceed any negative impacts it produces. Generally speaking, they're skeptical of the idea of precautionary regulation that tries to eliminate the possibility of risk or harm before real-world deployment even occurs. Instead, they argue that giving developers the space to operate as they see fit will produce the best outcomes fastest. They don't want government regulation. They don't want government support. They want a clear runway and complete autonomy to innovate.

Finally, there are the ***Bloomers***. Like the Zoomers, their perspective is fundamentally optimistic. They believe AI can accelerate human progress in countless domains. At the same time, they recognize that a technology as transformative and protean as AI cannot and should not be developed and deployed in a unilateral fashion. AI is going to impact too many lives in too many ways for that. So Bloomers pursue mass engagement, in real-world conditions—which is what you get with iterative deployment. While they're not unconditionally opposed to government regulation, they believe that the fastest and most effective way to develop safer, more equitable, and more useful AI tools is to make them accessible to a diverse range of users with different values and intentions.

Instead of seeing AI as fundamentally an extractive industry, as many Gloomers do, Bloomers see it as more akin to agriculture. You plant a seed. You watch it grow and adapt to the given conditions. You learn what crops work best where, and begin to intervene in ways informed by everything you learn about the problems and challenges that arise and the solutions that can potentially mitigate them. It's not a risk-free process. Over time, though, your knowledge increases, your techniques improve, your yield grows.

If you haven't guessed already, I place myself in the Bloomer camp. From my role in launching LinkedIn, to investing in hundreds of internet startups over the last twenty-five years, I've built my career on the principles of network feedback loops, and learning from and making improvements based on real-world usage.

It's the emphasis on broad participation in self-directed ways that draws me toward the Bloomer perspective, because participation both requires and rewards individual agency. In contrast, both Doomers and Gloomers presume to protect human agency—by banning or restricting AI, and thus taking away our capacity to assess it ourselves. With Zoomers, it's the other way around: they want absolute freedom to build, generally under the assumption that the outcomes will ultimately enhance human agency. But this perspective often fails to consider how their preferred breakneck pace makes the rest of humanity feel in terms of their stake in the process.

By enabling people to use AI tools in hands-on ways, OpenAI instantly broadened the conversation around our potential AI future and made it more inclusive. This approach is also giving everyone more information to evaluate the opportunities and challenges these tools create.

As exciting as the tech breakthroughs driving AI over the last few years have been, this democratization of access has been equally exciting, and arguably more important. If we are in fact in the midst of a transition on par with the advent of steam power, then the decisions we make now won't just shape the next few years. They'll shape the next few decades, and even the next few centuries. In the age of networks and digital empowerment, no one should be relegated to the role of nonplaying character in the creation of their own future. To develop

AI democratically, responsibly, ethically, and efficiently, we must incorporate broad public participation and continuous public feedback. Humanity has entered the chat, and that's a hugely necessary plot development.

There's a strategic global element to this aspect of the story too, and it ties back to individual agency. Because ultimately a positive AI future isn't just about which companies or even countries develop the technology the fastest. Outcomes will also depend on which individuals, companies, and countries use AI most productively. Which is another way of saying the key issue here is trust.

As developers introduce more capable models, especially ones that can function more autonomously, with minimal human oversight or interaction, the cultural clashes these AIs are already inspiring will only increase. The desire to preemptively regulate and even prohibit them will grow stronger. But all of this is playing out on a global scale. Any regulatory decisions the U.S. or any other country makes won't manifest in a vacuum. America's AI future will in part be determined by decisions that China, the European Union, India, and dozens of other countries make, and vice versa. It's very much a multiplayer game.

In light of this, discussions on AI adoption often coalesce around regulation. Zoomers demand the freedom to innovate. Gloomers advocate for pauses, prohibition, and top-down licensing regimes. Bloomers believe the freedom to innovate and the obligation to regulate are both important, and that we must strike the right balance between the two. In part, this means recognizing that people have widely divergent views on AI, and that our primary goal should be the pursuit of broad consensus and trust. And this is something that is most likely to emerge out of access and use. In other words, it's not enough to make rules about AI. Beliefs, norms, and shared values matter too.

As we'll explore in the chapters ahead, AI is a technology that can dramatically empower individual users while also functioning on very broad societal levels. In this way, it's akin to the internet itself, or roads and highways, or the electrical grid, or the water system. To varying degrees, for varying reasons, there are people who opt out of all these systems, or at least live without them. But imagine if, say, 30 percent of the public had deep reservations about power lines. Or 20 percent were adamant about not incorporating traffic lights into their lives. Finally, say a very tiny percentage of the public, no more than a few thousand people, or even a few hundred, hated reservoirs so much they spent a fair amount of their time actively trying to destroy or undermine them.

None of this behavior would likely shift our commitments to the utilities and infrastructure that enable us to live rewarding twenty-first-century lives. But it likely would have impacts on how smoothly and productively we, as a country, function, with both domestic and global implications. That's why it's crucial to develop AI in ways that give the public a real role in the process, just like we did when developing automobiles and the internet. Obviously, uniform consensus is both unlikely and unnecessary. But with so much at stake, it makes sense to proceed in ways that give us the best chance of success.

In large part that's because a perverse aspect of governance in an era of digital networks is that societies where democratic input is broadest are also those where change becomes hardest to make. Cacophony trumps community and consensus. So that's the special challenge that faces the U.S. and other democracies now. If we falter, there's a very real chance more authoritarian, or at least more socially cohesive, countries will adopt and deploy AI technologies more effectively than we do.

Think of all the ways that could undermine individual agency. We already know that AI tools can be designed to work on you rather than for you or with you, even in democracies. In autocracies, that likelihood increases. Also consider national prosperity and national security. Protectionism won't allow us to economically compete with countries equipped with millions of highly intelligent robot workers and an even greater number of AI systems functioning as synthetic scientists and engineers. Isolationism won't protect us from the exotic new autonomous weapons those synthetic scientists and engineers might devise.

All of this means that in the twenty-first century, individual agency is more closely aligned with national agency than ever before. To ensure an AI future that functions as "an extension of individual human wills," democracies must lead this effort. That means developing with an ambitious innovation mindset, a commitment to experimentation and adaptation, and broad public participation to build consensus along the way.

With that in mind, let's move forward by taking a step back and looking at another moment in time when computers were on the verge of changing everything.

CHAPTER 2

BIG KNOWLEDGE

When George Orwell published *1984* in 1949, the entire world had fewer computers than you can find now in a busy Starbucks. Even TV networks were still in their infancy. So, Orwell's take on how electronic audiovisual networks would come to shape our lives so profoundly was obviously very prescient, and helps explain why *1984* still continues to show up on Amazon bestseller lists every few years or so.

1984 takes place in a country formerly known as England, now dubbed Airstrip One. It's the third most populous province of Oceania, a larger superstate governed by a regime known only as the Party and its omnipresent figurehead, Big Brother, who may in fact be a deepfake. The Party brutally subjugates its citizens through a soul-crushing combination of constant surveillance, frequent torture, and propaganda so absurd it extinguishes any impulse toward logic, dignity, or hope. Once you've bought into slogans like IGNORANCE IS STRENGTH and FREEDOM IS SLAVERY, what else is left except total compliance?

In Oceania, everything is gray and smeared—the streets and skies, the past and the future, truth itself. Only the posters of the country's mustachioed leader provide a hint of dynamism and color, and even they are darkly foreboding. BIG BROTHER IS WATCHING YOU, they declare, and definitely not in a reassuring I've-got-your-back way.

The state's omnipresence is accomplished through a comprehensive network of publicly and privately installed "telescreens" that continuously broadcast state propaganda while simultaneously surveilling Oceania's citizens. This network is overseen by an agency known as the Thought Police, who diligently scan for signs of disloyalty to Big Brother or the Party, along with pretty much any instance of human feeling, an interior life, or deviations from the norm. In Oceania, even the slightest indication of an interest beyond the bounds of the Party qualifies as noncompliance.

In *1984*, Orwell presented such a harrowing vision of God-level techno-surveillance and its dehumanizing effects that we've never been able to quite unsee it. For more than seventy-five years now, multiple generations of Gloomers have been prophesying an apocalypse that's just around the corner.

In the early 1960s, the growing prominence of mechanical behemoths known as mainframe computers led to a pronounced spike in such doomcasting. By 1963, the U.S. computer industry had manufactured around 15,500 computers.[1] About a thousand of them were leased or owned by the U.S. government, and the rest found homes at large corporations and universities, where they crunched numbers at groundbreaking speeds and gave these institutions powers of perception and insight that had previously been impossible.

Case in point: When the Internal Revenue Service announced in 1963 that it was putting a $1.2 million IBM 7074 into service to process and verify tax returns that year, worried taxpayers streamed into IRS offices to make good on previous underpayments. All told, an IRS official told the magazine *Popular Science*, the agency collected $700,000 in unpaid taxes.[2]

Even at this early stage of development, the breadth of what computers were accomplishing was impressive. News organizations were using them to predict election results. Ice cream manufacturers were using them to precisely control their blending operations. New York City was using them to manage traffic-signal timing patterns. Astrid Lindgren, the Swedish author of the Pippi Longstocking books, received an additional 9,000 kronor, or approximately $11,700 in 2024 U.S. dollars, in royalties from the Swedish public library system, where an author compensation system for public lending had been put in place in 1954.[3] "Since this [payment] was based on 850,000 total loans of her books from thousands of schools and libraries, the bookkeeping was possible only with an electronic computer," D. S. Halacy Jr. wrote in his 1962 book, *Computers: The Machines We Think With*.

Of course, what might have been described as a new Light Ages, where institutions of all kinds were utilizing mainframes to collect more data, obtain new clarity about their operations, and take actions based on this more comprehensive knowledge, had a dark edge too. "There are banks of giant memory machines that conceivably could recall in a few seconds every pertinent action—including failures, embarrassments, or possibly incriminating acts—from the lifetime of each citizen," Vance Packard, one of the most widely read writers in his era, exclaimed in his 1964 bestseller, *The Naked Society*. "And brain research has progressed to the point where it is all too readily believable that a Big Brother could implant an electrode in the brain of each baby at birth and thereafter maintain by remote control a certain degree of restraint over the individual's moods and behavior, at least until his personality had suitably jelled."[4]

It was even more probable—though decidedly less alarming—that the giant memory machines might help economists analyze public health data in ways that could reveal complex relationships between economic factors, health care utilization, and the health outcomes for various populations. In April 1965, the Committee on the Preservation and Use of Economic Data, a group of social scientists that had been working at the behest of the federal government for several years, published a report that proposed that the government establish a "Federal Data Center" that could provide more centralized and easier access to the growing amounts of data that the government's various agencies were now accumulating.

The basic idea was to gather a wide range of data that twenty different federal agencies had collected, and make it more accessible and usable to researchers. This, in turn, could help inform federal policymaking at a time when the government was beginning to expand social programs as part of President Lyndon B. Johnson's Great Society vision. But at that point, the data that the government had begun to amass was siloed in more than 600 datasets. These datasets were stored, in highly decentralized and often-incompatible fashion, on more than 100 million punch cards and 30,000 magnetic tapes. Consolidating this data in a single physical location—a national data center—the social scientists reasoned, would make the data more useful and therefore create more value for the taxpayers who'd underwritten its collection.[5]

The committee's timing couldn't have been worse. Along with the growing prevalence of mainframe computers, technological breakthroughs in transistor and printed-circuit design had enabled manufacturers to make electronic devices simultaneously smaller and more powerful. Mail-order catalogs advertised in men's adventure magazines offered microphones the size of sugar cubes and closed-circuit TV cameras that could be hidden in a vest pocket. Private investigators, aspiring corporate spies, and the plain nosy could purchase parabolic microphones that could pick up conversations a city block away, voice-activated tape recorders, and infrared cameras that could take photos in total darkness.

All of this, midcentury Doomers and Gloomers postulated, pointed toward the hyperinstrumented world of Airstrip One and Oceania. "[The] most frightening implication for the future is the notion of built-in surveillance—that is, surveillance systems built directly into new hotels, schools, jails, office buildings or other structures," wrote Myron Brenton in *The Privacy Invaders*. "In this Orwellian nightmare, the wiring could lead to a central listening and recording station."[6]

Along with growing public concerns over new kinds of electronic snooping, the Committee on the Preservation and Use of Economic Data also had *Griswold v. Connecticut* to contend with. Just two months after these social scientists had published their report, the Supreme Court struck down a Connecticut statute prohibiting the use of contraceptives. As the historian Sarah E. Igo suggests in *The Known Citizen: A History of Privacy in Modern America*, the court's ruling in *Griswold v. Connecticut* "identified a constitutional 'right to privacy' for the first time in American history."[7] In addition, it established "the potential to generate new privacy claims and also sidestep others."[8]

So it was definitely not a great time to be publicizing the fact that the U.S. government had compiled the "the most comprehensive statistics program in the world" and had ambitious plans to make all that information more accessible to egghead technocrats.[9] When the Johnson administration announced it was moving forward with this instance of Great Society optimism and good governance, the backlash from Congress was "swift and strong."[10] A number of senators and members of Congress had already been holding hearings addressing the privacy-related issues that had been raised in books like Packard's and Brenton's. They quickly added one more to their calendars.

In a July 1966 subcommittee hearing addressing the topic of "The Computer and Invasion of Privacy," lawmakers acknowledged that they weren't necessarily opposed to what the national data center was intended for—using actual data to inform government policies and laws—but they were worried about what such a center might become. "It is our contention that if safeguards are not built into such a facility, it could lead to the creation of what I call 'The Computerized Man,'" warned Cornelius E. Gallagher, a Democratic member of Congress who represented New Jersey's 13th district. "'The Computerized Man,' as I see him, would be stripped of his individuality and privacy. Through the standardization ushered in by technological advance, his status in society would be measured by the computer, and he would lose his personal identity. His life, his talent, and his earning capacity would be reduced to a tape with very few alternatives available."[11]

"There is danger that computers, because they are machines, will treat us as machines," echoed Frank Horton, a Republican representative from New York. "They can supply the facts and, in effect, direct us from birth to death. They can 'pigeonhole' us as their tapes decree, selecting, within a narrow range, the schooling we get, the jobs we work at, the money we can earn and even the girl we marry."[12]

Vance Packard himself made a trip to the newly completed Rayburn House Office Building on Capitol Hill to educate the subcommittee on how both government and private-sector "filekeepers" had amassed billions of records on millions of U.S. citizens, filled with "derogatory information of one sort or another"[13] about every aspect of their lives.

Through tax returns, job-related personality assessments, lie-detector tests, medical records, credit reports, insurance company investigations, police records, and even moving company inventories, Packard suggested, authorities could compile dossiers that might include information about people's health conditions and emotional well-being, their sex lives, the state of their marriages, their brushes with the law, and their known associates and affiliations.

In Packard's estimation, the creation of "one giant system [that could] be activated for purposes of inserting or retrieving information from numerous locations" would ultimately lead to the "depersonalization of the American way of life" and "a suffocating sense of surveillance as the public learns that their Government is developing an all-seeing eye."[14]

"Let us remember, 1984 is only eighteen years away," Packard noted as he wrapped up his testimony. "My own hunch is that Big Brother, if he ever comes to these United States, may turn out to be not a greedy power seeker, but rather a relentless bureaucrat obsessed with efficiency. And he, more than the simple power seeker, could lead us to that ultimate of horrors, a humanity in chains of plastic tape."

News media coverage of the hearing and the proposal for a national data center echoed such sentiments. The *Charleston Post* worried that greatly enhanced government records would result in dossiers like those compiled by "Hitler's Gestapo," only worse. The *Benton Harbor News-Palladium* suggested elected officials might use it to coerce campaign contributions from individual citizens or blackmail political opponents. "Can personal privacy survive the ceaseless advances of the technological juggernaut?" a *New York Times* editorial lamented, before invoking "the Orwellian nightmare" that would descend upon America if Congress gave those twenty federal agencies permission to organize their data more efficiently.[15]

For several more years, members of Congress continued to use the idea of a national data center as an ideological punching bag in hearings and reports. Then, in 1969, the initiative was finally put to rest as a result of the bureaucratic reshuffling that commenced when Lyndon

Johnson's term ended and the stewardship of America's data future passed to Richard Nixon. The latter, of course, would go on to prove himself to be a surreptitious file keeper of some renown.

These days, Vance Packard is mostly forgotten. The national data center and the howls of protest it inspired from Congress, op-ed writers, and the American public are all but unknown. In part, that's because even in the short term, the defeat of the national data center did little to inhibit the pace of technological innovation. First, hard disk drives replaced the slow and limited-capacity magnetic tapes of midcentury mainframes. Then we got time-sharing systems, minicomputers, local area networks, the PC revolution, video cameras, the World Wide Web, social media, the cloud, mobile phones, and fitness trackers. And also motion sensors, facial recognition systems, fingerprint scanners, eyeball readers, smart thermostats, sleep trackers, RFID implants, digital tattoos, and data brokers whose digital profiles of you potentially include thousands of data points about your basic demographics, medical conditions, financial status, sexual preferences, political leanings, and even predictions about your future behavior—all compiled without your direct knowledge or consent.

In many respects, the apprehensions those mid-century Doomers and Gloomers expressed about individual agency, privacy, and the all-seeing eye of technology retain their currency. Indeed, it's easy to imagine Packard posting very concerned essays on his Substack, regarding an LLM's capacity to manipulate public behavior through the scale production of persuasive content tailored to exploit individual psychological profiles. Or Cornelius Gallagher time-traveling from 1966 through some wormhole in the Capitol Rotunda to grill Google's Sundar Pichai about algorithmic decision-making.

But it's also worth noting how colossally wrong about the future Packard and company were.

Big Brother Is Making You Feel Seen

When Packard warned Congress about a world where technological advances would leave "humanity in chains," many states still had laws restricting interracial marriage. Same-sex marriage was so inconceivable no state had even yet thought to pass laws explicitly prohibiting it. *Griswold v. Connecticut* didn't establish a right to privacy for everyone who might want to use contraception. Justice William O. Douglas's majority opinion repeatedly notes that the court's ruling specifically protects "the right of marital privacy" only. In a concurring opinion, Justice Arthur Goldberg made it even clearer that the court's decision was not intended to infringe on "a State's proper regulation of sexual promiscuity or misconduct" or deny it its power "to forbid extramarital sexuality . . . or to say who may marry."[16]

So while opponents of the national data center believed they were making a principled stand for American liberty and freedom, many of their fellow citizens were very much encumbered by legislative chains of one kind or another.

But it's also true that cultural norms were shifting. And as computer systems proliferated, we ended up getting a much freer world than the one that preceded it. This new world championed creative exploration, rebellion against conformity, alternative lifestyles and values. It extended individual agency and autonomy to traditionally marginalized groups, including women, people of color, and the LGBTQ community.

Many factors drove these changes—especially grassroots organizing and activism that led to new laws and regulations. But technological innovation played a key role too. Notwithstanding all those magazine ads for hidden microphones and infrared spy cameras, the late twentieth century was shaped more by slightly different devices: remote controls, cable TV, VCRs, Sony Walkmans, and handheld video cameras.

Each chipped away at the conformity of the mass-media era and helped usher in a more personalized, configurable, and self-actualizable future.

Most remarkably, the mainframe computer, that terrifying apparatus of creeping totalitarianism, was largely superseded by a new machine that might have struck Vance Packard as implausibly oxymoronic: the *personal* computer.

"What kind of man owns his own computer?" an early Apple ad asked—and answered with an illustration of creative-class icon and exemplar of self-improvement and individual autonomy Benjamin Franklin. "Rather revolutionary, the idea of owning your own personal computer?" the ad continued. "Not if you're a diplomat, printer, scientist, inventor . . . or a kite designer too. Today there's Apple Computer. It's designed to be a *personal* computer. To uncomplicate your life. And make you more effective."[17]

But it wasn't just that new tools like the computer were giving individuals new capabilities and new levels of agency in their personal lives and their work lives. While Packard was prescient in noting how corporations, along with the government, had begun to amass more information about individuals, this practice didn't dramatically constrain the options and alternatives available to people as Packard and others had feared.

Instead, the giant memory machines that Packard warned would oppress future generations of Americans helped liberate them. The new capacity to aggregate and analyze vast amounts of consumer data gave manufacturers, marketers, and content creators the power to segment markets, target more granular demographics, and cater to specific subcultures, lifestyles, and interests in ways that could never happen in the pre-computer age. Personalization across multiple domains began to replace the standard one-size-fits-all approach of twentieth-century industrialism.

It was the beginning of a more diverse and inclusive world based on comprehensive knowledge. Big Brother, or, perhaps more appropriately, Big Business, wasn't watching you so much as making you feel seen. Equipped with a greater awareness of what people were buying, what information they were consuming, and generally how they were living their lives, commercial enterprises could serve them more effectively.

In today's highly digitized and networked world, what was once opaque becomes easy to analyze—there's no question about that. But what's an infringement on privacy and what's an assertion of identity? When does the production and dissemination of new knowledge bolster individual agency and when does it diminish it? How much does an allegiance to privacy, to withhold information and keep it close to the vest, reduce our opportunities to share knowledge, learn new things, grow?

Everyone has "a right to be left alone,"[18] as future Supreme Court justice Louis Brandeis put it in an 1890 *Harvard Law Review* essay that helped shape twentieth-century conceptions of privacy. But as a member of society, and especially as a member of a networked society, it's not always the most productive right to organize your life around.

In the 1960s, when *1984* still cast such a long shadow and the accelerating pace of change was so disorienting, it was seemingly impossible to envision a world where devices capable of capturing, analyzing, and sharing information could enhance human agency instead of diminish it.

For some, it still is. You don't have to look far on today's social media platforms to find progressive privacy hawks or right-wing truth-tweakers lamenting Big Tech's Orwellian overreach online, and its suffocating impact on privacy, personal autonomy, and twenty-first-century culture in general. But really, who today has *fewer* options than they might have had in the 1960s, or the 1980s, or the 2000s to express who they are, prioritize what they value, and pursue a life of their own making?

For all practical purposes, Big Brother has been dead for decades. If that sounds hyperbolic, just try to imagine the overwhelm that Winston Smith, *1984*'s protagonist, would experience if you dropped him in the streets of contemporary Manhattan. The kaleidoscopic storefronts alone might knock him into an uncomprehending stupor. Then add the dizzying presentations of self, the bold displays of personal intimacies and enthusiasms that have nothing to do with the ruling party. The giant billboards filled with images of people who aren't Big Brother. The smartphones that people clutch to their hearts like cherished talismans.

In the U.S. today, our current environment is so geared toward self-expression, diversity, and nonconformity that a large swath of the public apparently yearns for the resurrection of a Big Brother–like strongman who can inspire absolute loyalty and restore social cohesion to our polyvocal and highly balkanized republic.

And yet, because of AI and its appetite for data, many of today's critics are more sure than ever that the "chains of plastic tape" that Packard prophesied sixty years ago may finally ensnare us. Certainly, it's not a possibility we should dismiss lightly. Past performance is no guarantee of future performance, especially in the disruption business. But just as we shouldn't necessarily discount the failed predictions of the past, we shouldn't discount its actual lessons. And what history has shown for the past sixty years, over and over, is that information, broadly shared, collected, analyzed, and deployed in various ways, can be incredibly empowering and uplifting—not chains of plastic tape so much as climbing ropes you use to scale greater heights and achieve new levels of meaning and fulfillment.

This doesn't mean that privacy isn't valuable. Especially in the last three hundred years or so, after urbanization began to really kick in and large numbers of people started living in close proximity to strangers, we've considered privacy increasingly sacrosanct. Outside the pressures, pieties, and often-restrictive norms of public life, it gives us space for candid self-reflection and self-assessment. It enables us to detect and define our authentic selves. That in turn helps us act more autonomously, with more capacity for agency.

But privacy isn't the only way to achieve these ends. Especially in a networked world, a strong public identity creates autonomy and agency too. Even if we sometimes seek refuge from the perpetual disclosures of digital life via pseudonyms, VPNs, and encryption, we also want the validation, acceptance, and personal sovereignty that public identity facilitates. That's why a signature aspiration of the Digital Age—maybe *the* signature aspiration—is to be seen.

Public identity equates to discoverability, trustworthiness, influence, power, agency. It's a form of social capital—and sometimes financial capital—that can help you move more productively through the world.

This is what opponents of the national data center project failed to see sixty years ago. It's why contemporary Gloomers maintain the sacred traditions of their tribe and insist that, as always, we're on the brink of Orwellian dystopia, even as technological innovation continues to endow us with new superpowers that enable us to live more meaningful and fulfilling lives.

Introducing a Network of Trust

There's a reason OpenAI's founding vision of AI as "an extension of individual human wills" struck a chord in me. That same desire to create new ways for people to attain more agency in their lives—the power to make deliberate choices and productively act on them—is what initially inspired LinkedIn. As we explained on the website circa 2003, the year we launched, "You're already connected to the people you need to reach your business goals—through the business connections you already have. Make contact with thousands of professionals, through trusted network connections, and help yourself and your colleagues get it done."[19]

Still, in its early years, LinkedIn was often simply described as a "résumé site." Or a "job listings site." In fact, twenty-plus years on, that's how some people still think of it. From the start, though, I always envisioned LinkedIn as something more foundational than that.

For me, LinkedIn was always about using networks to share and discover information in new ways, by using identity to increase trust. Creating a publicly accessible profile that includes substantive information about your job and career aspirations may seem obvious now, but in the early 2000s it was not yet default behavior. In fact, when the internet was just starting to go mainstream with the introduction of the World Wide Web a decade earlier, anonymity had been cast as one of its great virtues. As a classic *New Yorker* cartoon from 1993 put it, "On the internet, nobody knows you're a dog."*

But what if you *wanted* people to know you were a dog? Or more to the point, what if you wanted people to know you were a database engineer with deep proficiency in MySQL and SQL Server? And be easily discoverable as such? And better yet, be easily trustworthy as such?

Prior to the internet, pretty much only celebrities and executives

* Created by Peter Steiner, this single-panel cartoon was originally published in the *New Yorker* on July 5, 1993, and went on to become the most reprinted cartoon in the magazine's history. In October 2023, Steiner's original drawing went up for auction for the first time. An anonymous bidder—naturally!—paid $175,000 for it, the highest amount ever for a single-panel cartoon.

who were prominent enough to be covered by magazines like *Fortune* or the business section of the *New York Times* had a professional identity that carried beyond their coworkers and limited circle of people they did business with directly.

With the emergence of the web, anyone could create and publicize a professional identity for themselves. This was a major shift with many potential upsides, but it also created trust issues. Without *Fortune* or the *New York Times* on hand to certify your assertions about yourself, how could you assure people that you really were a database engineer with deep proficiency in MySQL, and not just a dog pretending to be one? Situating individual identity within a larger network of professional connections is one way to do that. You might not know Mark, but if you know Linda, and you know Linda knows Mark, then Mark instantly seems a little more trustworthy.

At heart, then, what LinkedIn and many other successful internet platforms do is scale trust. Think about how eBay, PayPal, Airbnb, Uber, and Lyft, to name just a few, used various innovative trust mechanisms to enable a broad new range of interactions, transactions, and behaviors on a global basis.

While the internet has certainly created new opportunities for fraud and disinformation, its larger story is how it has functioned as an unprecedented trust machine. In 1995, for example, we were net-surfing under the cover of pseudonyms. By 1999, we were buying used cars on eBay Motors, sight unseen. By 2012, we were jumping into a random Toyota Corolla with a pink mustache on its grill after a night on the town to get a safe ride home. Pretty amazing when you think about it.

In the case of LinkedIn, as its user base grew and people's individual networks got bigger and more densely connected, the platform became increasingly trustworthy. That's because every user was effectively vouching for many other users, in a new kind of distributed trust platform rooted in real-life individual identity.

To make this new kind of trust platform work, though, we had to convince people to share more information about themselves than they were used to sharing online. And it wasn't just that we asked our users to create profiles for themselves using their real names, which in itself was still a fairly uncommon practice at the time. We also encouraged them to share their job titles, the companies they worked for, their skills and experience, and their networks. To be sure, we gave users explicit privacy controls that enabled them to manage who could see their information and how it was shared. But generally, the idea was to make yourself at least partially visible to other LinkedIn members beyond your immediate connections, in order to facilitate new connections.

In 2003, this was definitely not business as usual. In fact, a lot of individuals and companies found it downright controversial. "LinkedIn. Controversial?" you may be saying to yourself right now. "How are those two words even within a hundred yards of each other?"

Today, LinkedIn has more than a billion global members. Creating a profile page for yourself, with a detailed record of your professional aspirations and accomplishments, *is* business as usual. Before the internet, though, one reason that people generally didn't share detailed information about their careers more broadly was simply that there were few if any ways to do so. You told people about your experience and expertise when you were actively looking for a new job. Or maybe when you were chatting with a stranger at an airport bar while you were waiting for your flight to start boarding. Résumés were sent directly to company HR departments, professional recruiters, and employment agencies.

In the desktop-PC era of the 1980s and early 1990s, some people used applications like ACT! Contact Manager and Sidekick to keep track of their professional networks—but these weren't networks in the way we understand that term today. Instead, you maintained a database on your own computer of people's names, phone numbers, email addresses, employer, and whatever other information you'd managed to glean about them. When their information changed, you'd have to update it manually in your own file—provided you knew that the change had occurred.

By the early 2000s, some developers had created web-based contact management applications. But these systems simply replicated the approach that the desktop applications had used. So, while users could now access their own lists of contacts from any computer with an internet connection, all of these lists were effectively still their own tiny islands of data, disconnected from everyone else's. Without really meaning to, this approach maximized privacy—but it also failed to take advantage of all the valuable new functionality that a genuine network could deliver.

That's what LinkedIn did. On our platform, users connected with other users, and created and maintained their own profiles. You no longer had to check in periodically with your old boss from your last job to see if your contact information for her was still current, and update her record in your database if it had changed in some way. Instead, she updated her LinkedIn profile when anything changed, and anyone that she was connected to would immediately have access to her most current information.

Solving the update problem was just the start of the benefits of connecting information in this shared and more public manner. First and foremost was the fact that you were now in charge of your own personal identity. Instead of other users creating an entry for you in their contact lists and including what they knew about you, you presented the information that you wanted them to know. You were in charge of defining yourself.

Once you connected with someone else, you could also now see their contacts, something that had never been possible in the desktop-contact-app era. Now it was suddenly worlds easier to find new contacts and make yourself discoverable to others.

The challenge we faced was that these were all new behaviors. When LinkedIn launched, real life and "cyberspace" were beginning to converge in more ways, but anonymity or pseudonymity were still the reigning modes for public interaction on the internet. And people were especially wary about sharing details about their professional lives.

In fact, in our early days, even as social networks were starting to get popular, most people felt they had clear incentives *not* to create profiles on LinkedIn, precisely because we asked our members to use their real names. Some people didn't want their bosses to think they were unhappy and about to jump ship. Others didn't see the point of publicly sharing real information about themselves if they weren't currently looking for a job or seeking some other near-term payoff.

Finally, if you were the kind of person who actively compiled a professional network of contacts—which at the time meant there was a pretty good chance you worked in sales—your network was one of your great competitive advantages. So why share it with potential competitors? Why open yourself up to a new stream of digital cold-callers pestering you for easy introductions to the people you knew?

Employers weren't necessarily crazy about the platform either. Because even if their employees simply wanted to network in ways that would help them do their current jobs better, why allow them to make themselves so visible to potential recruiters and competitors? And not merely visible, but free to describe their jobs as *they* saw them, which was perhaps not quite how their employers did.

Fortunately, enough early adopters felt, as we did, that the many potential upsides our platform could eventually deliver made participation an intelligent risk worth taking. By sharing information about themselves, they would be able to increase the utility and scope of their own professional networks. This in turn could provide a wide range of benefits over time: new opportunities for professional identity development, new sources of industry intelligence, and new potential partners to collaborate with in different ways.

We Don't Know What We Know

Had something like LinkedIn existed in the 1960s, Vance Packard would have probably found it unsettling. It is, after all, not just a national data center but a global one. It amasses a huge amount of information into a single repository and makes it easy to search. Much of this information would no doubt strike Packard as fundamentally "private," and thus problematic, even if posted voluntarily. (Much of the data that the government and private-sector players aggregated in the 1960s was voluntarily disclosed as well.)

Yet this willingness to live more publicly creates a great deal of collective value. With so many other people posting details about their own work histories and aspirations, you can learn what kinds of skills are important to the people who have the job you'd like to have, at the company you'd like to work at. You can see which companies are most popular with other applicants.

Once that formerly siloed information is easier to access, everyone participating on the platform, including individuals and companies alike, can operate in more knowledgeable ways. To illustrate just one way this plays out: every minute, seven people are hired via LinkedIn, a rate that equates to 3.67 million people a year.[20]

It's also true that the more public lives we lead now don't always play out in beneficial ways. A casual, supposedly ephemeral exchange on Facebook or X.com may now stay public indefinitely. Screen-capped or retweeted by digital bystanders, an off-the-cuff remark can become vastly more public than its writer ever intended. In this new reality, it makes sense that the refuge that privacy affords grows even more valuable to us.

Yet societies—and especially democracies—have always depended on the free flow of information at least as much as they depend on privacy. In their 2018 book, *Network Publicy Governance,* internet theorists Andréa Belliger and David J. Krieger write, "It would seem that we are entering the 21st Century with society divided into those who believe that as much information as possible should be integrated in all areas and at all levels and those who believe that human freedom, autonomy, and even dignity depend on secrecy and the withholding of as much information as possible."[21]

How can we strike the right balance between these two viewpoints? In the face of major computing innovations in the 1960s, American lawmakers and pundits focused almost exclusively on what could possibly go wrong, at least in the case of the national data center. In their successful efforts to prohibit that project, they also prohibited whatever benefits might have come from making the federal government's growing stores of data more accessible to researchers and policymakers, sooner rather than later.

What's the hurry? The more we embrace AI, the more effectively we can tap into the data and information we're already creating in ways that will increase our agency across nearly everything we do as humans. This is because humanity has already reached the point where we're producing more information than we can effectively make use of on our own. In a 1998 essay, Hal Varian, an economist and professor at the University of California, Berkeley's School of Information who would eventually become Google's chief economist, wrote how the amount of information that the world was producing that was actually consumed was "asymptoting towards zero."[22]

To support this claim, Varian cited *Communications Flows: A Census in the United States and Japan*, a 1984 book coauthored by Ithiel de Sola Pool, a longtime professor at the Massachusetts Institute of Technology. In it, Pool and his coauthors attempted to quantify the actual data underlying the "information explosion." To do this, they tabulated the total number of "words supplied" and "words consumed" each day in the United States and Japan for a given year, from eighteen different media sources, including radio, TV, books, newspapers, magazines, and what was at the time a fairly new but fast-growing category, "data communications." In the U.S., they estimated that "words supplied" grew from 8.7 million each day in 1975 to 11 million each day in 1980. In contrast, the "words consumed" each day for those years went from 45,000 to 48,000.[23]

You can sense how this played out over time. On one side of the ledger, there was an expanding array of radio stations, cable channels, catalog retailers, CD-ROM publishers, internet newsgroups, email spammers, streaming video platforms, podcasters, and TikTok influencers supplying millions and millions more words per day. And on the other side of the ledger, there was our slow, easily distracted, time-constrained brains, which probably maxed out on "words consumed" in the early 1990s or so.

By Pool's estimates, the average U.S. inhabitant was already only consuming 0.004 of the information available each day in 1980. And today? In the time it takes you to read this sentence, the world now produces enough data and information to fill 23 billion e-books.* Some of that comes from humans, in the form of tweets, Wikipedia entries, GitHub repos, white papers posted to arXiv.org, IRS guidances, and TikTok dances. Some of it comes from smartphones, smart thermostats, security cameras, and other IoT infostructure, in the form of GPS data, temperature readings, video footage, and more. What this means, of course, is that even if you catch every episode of Kara Swisher's podcast and read most of Matt Yglesias's tweets, you're still basically living in your own personal Dark Ages, ignorant of virtually all global knowledge.

And that's precisely why AI is so crucial to our future, as individuals and collectively. In the same way we used steam power, and now use multiple sources of synthetic manpower, to amplify human agency, we can now use AI. Distributing intelligence broadly, empowering people with AI tools that function as an extension of individual human wills, we can convert Big Data into Big Knowledge, to achieve a new Light Ages of data-driven clarity and growth.

---

* According to the Worldwide IDC Global DataSphere Forecast, 2023–27, worldwide data production in 2024 is estimated at 147 zettabytes, or roughly 402 billion gigabytes per day. Given that a typical e-book requires two megabytes of storage, that equates to enough data to fill roughly 201 trillion e-books a day, or 139 billion books per minute, or 2.3 billion books per second. If it took takes you ten seconds to read that sentence, the world produced enough data to fill 23 billion e-books in that time. See John Rydning, Worldwide IDC Global DataSphere Forecast (International Data Corporation, 2023).

CHAPTER 3

WHAT COULD POSSIBLY GO RIGHT?

A new Light Ages? Because these days you can even buy AI toothbrushes that optimize your brushing style through real-time coaching? Like many Silicon Valley founders and investors, I sometimes get accused of solutionism. That's the belief that holds that even society's most vexing challenges, including those involving deep political, economic, and cultural inequities, have a simplistic technological fix.

As we suggested in the last chapter when discussing the cultural shifts that informed the 1960s and the decades that followed, government legislation and collective action typically drive social progress. In part, that's why we emphasize OpenAI's iterative deployment approach as a general strategy for AI development. It's a form of collective action that gives the public an opportunity to provide ongoing input into deliberately incremental versions of the product.

But technology is itself one of humanity's most proven levers for creating positive change at scale. That's why solutionism's inverse, *problemism,* is a real issue we face too. Problemism is the default mode of the Gloomer who views technology in general as a suspect force, antihuman and antihumane, tinged with the new-car reek of capitalism.

Die-hard Gloomers typically cast themselves as watchdogs for the public good, ethical counterforces to solutionism's trickle-down immiseration. In doing so, they can in fact be valuable in holding Big Tech and other powerful interests accountable for missteps, oversights, and other failures that create harmful outcomes for society.

But in emphasizing critique over action, precaution and even prohibition over innovation, problemism can also do real harm to society. When you only focus on what could possibly go wrong, you inevitably discount what could possibly go right. In the case of AI, it's true that there are harms it may cause, flaws we should be working to correct, uncertainties we should be vigilant about.

But AI can also help us address our most pressing global challenges. Whether we're trying to make progress on sustainable energy production, health care, education, or cybersecurity, technology will invariably contribute 30 to 80 percent of any effective solution. This includes issues arising from AI itself. To combat deepfakes and disinformation, we'll need to use AI detection systems that can monitor platforms at scale and adapt to new challenges just as fast as the systems that are producing deliberately misleading content.

Thus, every effort we make to preempt an innovation's potential negative outcomes also preempts potential positive outcomes it may produce, whether those take the form of entirely novel capabilities or solutions to existing challenges and inequities. Think, for example, of an AI system that learns how to interpret and translate animal vocalizations, enabling humans to understand the needs of endangered species in ways never before possible and thus leading to more effective interventions to protect biodiversity. Or consider an AI system that devises a dramatically more efficient way to optimize supply chains for food distribution, significantly reducing waste and improving access to nutrition in food-insecure regions, thus improving health outcomes for millions of people a year.

Does it make sense to delay a tool that could assist and amplify the world's smartest scientists? Is it morally defensible to defer the development of a wide range of AI science tutors, AI health advisors, or AI legal assistants that can better enfranchise marginalized individuals and demographics? Should we sacrifice a leading role for the U.S. and democracy in creating the world's technological future?

While prudence, deliberation, and even skepticism are very necessary in the context of AI development, the ultimate goal is to make progress. That means that we must accept some level of risk and uncertainty so that we can take action and move forward. Innovation requires exploration. Anyone who says otherwise is not speaking truth to power so much as casting a vote for entrenched interests and inequities. In many instances, Gloomers operating in problemist mode are little more than champions of complacency.

This is why embracing a what-could-possibly-go-right mindset is so necessary for creating positive change. Thinking in terms of best possible outcomes doesn't mean you ignore potential negative outcomes. It means you avoid those outcomes by envisioning the future you want and steering toward it.

The Existential Threat of the Status Quo

From a problemist perspective, AI applied to mental health care is Category 5 solutionism, an algorithmic gust of tech-bro cluelessness speeding toward disaster at 157 mph. *Treating potentially delusional people in crisis with a system that hallucinates? What could possibly go wrong?*

Like the legendary sci-fi writer William Gibson wrote in his 1982 short story "Burning Chrome," however, the street finds its own uses for things.

"Why go to therapy when you can just talk to chatgpt?" wrote one X.com user shortly after ChatGPT's release in November 2022. "Boyfriend is now consulting ChatGPT for relationship advice in lieu of a therapist and I'm not mad about it," wrote another. "Everyone is talking about Google getting disrupted by ChatGPT. But what about THERAPY? Who wouldn't want a judge-free, empathetic therapist that is always there for you?" wrote a third.

It's possible even these optimistic takes underestimate how full this glass might be. Yes, using LLMs in mental health care may involve risks that no one should cavalierly dismiss. But mental health disorders and challenges have massive negative impacts on global well-being. An ongoing shortage of mental health professionals means that hundreds of millions of people go untreated; more automated services could dramatically expand access to care. It's also a field that is rooted in evidence-based practices (EBPs)—approaches that have been proven effective in rigorously designed clinical trials, but where traditional methods of collecting and analyzing data are labor-intensive, highly subjective, and very difficult to scale.

All told, AI can bring Big Knowledge to what many practitioners call the "black box of psychotherapy." It can assess millions of interactions across thousands of hours of transcripts, to see what EBPs are most effective in real-world contexts and which mechanisms of change actually drive positive clinical outcomes. It can enhance the agency of care providers by giving them new tools to assist them and instruct them. It can enhance the agency of care recipients by making mental health support more accessible and more personalized.

But when you view such possibilities through the lens of problemism, you prime yourself to see land mines rather than the green shoots that crop up as innovators attempt to break new ground. That's certainly what happened when tech developer Rob Morris posted a thread on X.com after his nonprofit briefly added GPT-3 functionality to its app. In January 2023, when the frenzy of hope, hype, and alarm around ChatGPT was really starting to crest, Morris explained how his team had added GPT-3-driven text generation capabilities to its peer-based mental health messaging service, Koko.

In a multi-tweet thread, Morris noted how users receiving messages that GPT-3 had authored or coauthored had rated them significantly higher than messages that Koko's human users had composed by themselves. Even so, he stated, he and his team made the decision to quickly remove the new option from its service. "Once people learned the messages were co-created by a machine, it didn't work," he wrote. "Simulated empathy feels weird, empty."[1]

Morris added some additional thoughts about the nature of AI-simulated empathy, but the tweet above is the one that made his thread go viral—and it's easy to see why. That's because the phrase "Once people learned the messages were co-created by a machine" makes it sound like Koko hadn't informed its users that some messages of support, ostensibly from other caring humans, were in fact being ghostwritten by a mathematical computation incapable of human feeling.

You don't have to be a data scientist to accurately predict what happened next. "YOU are trash," said one of the nicer responders. "I think you should be experimented on to see how human you are," noted another. "holy shit usually there has to be a big investigation to get tech bros to admit this type of immoral shitbaggery but you just offered it out there for free," exclaimed a third.

As angry X.com users rendered similar judgments and news media outlets began to cover the story, Morris tried to clarify the statements he'd made in his original thread. The "people" his tweet had referred to, he told Gizmodo, were not Koko's users but rather Morris himself and his team.[2] After a few days, as they began to sense which replies had been generated by GPT-3 alone, with no user modifications, they found them increasingly "sterile."

Morris also told Gizmodo and Vice.com that, during the time that GPT-3-assisted messaging was available to Koko's users, every message that had been authored or coauthored by Kokobot had included a disclaimer that it was "written in collaboration with kokobot."[3] (Kokobot is an AI-driven chatbot that functions as a kind of all-purpose host or guide within Koko's messaging service. More on that below.)

The backlash provoked by Morris's thread was obviously not completely unfounded. By design, Koko serves a vulnerable population—young people struggling with mental health challenges. And the tweet that Morris had composed was confusingly written, in a way that very much made it sound like Koko had not been operating in an ethical, transparent manner.

At the same time, it was equally clear that many if not most of the people decrying Morris and Koko had no real conception of how Koko operates, and made no attempt to learn more before assuming that Koko had misled and betrayed its users, that these users had suffered real harm, and that severe reckoning, karmic, legal, or both, would be coming Koko's way.

Had any of these outraged problemists tried out Koko's service, they would have found that it provides a simple interface for engaging in a peer-based, highly structured, evidence-based practice called cognitive reappraisal. While Koko was originally launched around a decade ago as a stand-alone iPhone app, it now works as a service on Discord, WhatsApp, and Telegram. If you're feeling anxious or stressed, you can use Koko to create an anonymous query that it shares with other Koko users. These users, in turn, can write short responses to your post.

As a component of cognitive behavioral therapy (CBT), cognitive reappraisal teaches you to modify the negative emotional responses you might have to a given incident or state of being by reframing your thoughts about it. If you're feeling overwhelmed by job demands, for example, you might try to reduce your anxiety by thinking, "Each challenge at work is an opportunity to learn and improve my skills."

A chatbot called Kokobot guides the service's users through the process of creating messages that incorporate principles of cognitive reappraisal. For someone requesting support and validation from other users (a *help requester*), this might mean nudging them to state a specific problem they're struggling with in a clear and declarative way. For someone providing support (a *help provider*), this might mean prompting them to incorporate empathetic phrases like "I've been there" or "Sorry to hear this" into their replies.

In addition to providing such guidance, Kokobot acts as an intermediary for every peer-to-peer interaction that occurs on Koko. So, you never actually see the usernames of other participants or interact with them directly; you're always interacting with Kokobot. Kokobot also analyzes and moderates message content in real time. If you type something that suggests you might be a danger to yourself or others, Kokobot will steer you toward help lines like Crisis Text Line or the 988 Suicide & Crisis Lifeline. If you write an off-topic or inappropriate reply to a user's help request, Kokobot won't show it to the user.

AI gives Kokobot these powers, but not generative AI. Instead, Kokobot relies on more traditional natural-language-processing and machine-learning techniques to analyze and respond to user inputs. Or at least it did until its developers briefly endowed it with some new functionality in the fall of 2022: an option that gave users the choice of letting Kokobot draft a complete reply to another user. Under the hood, this functionality was powered by GPT-3.

"Want to send some kind words? With Koko, or by yourself?" this version of the service asked. In the menu of choices that followed, one of them, Option B, was labeled "With Koko" and illustrated with a robot emoji. If a user chose that option, Kokobot would produce a response to a help requester's message, and you, as the help provider, could choose to either edit it further or send it to the help requester with no edits. Any messages that incorporated Kokobot's contributions that were sent to help requesters disclosed its participation. In other words, it was a fairly straightforward "copilot" approach to AI. And Kokobot's participation in the process was explicitly communicated every step of the way.

So while you could say the entire controversy was due to Morris's poorly worded tweet, that's not really true. Hundreds of people on X.com expressed their outrage on X.com on behalf of Koko's misled and potentially harmed users. Multiple news outlets covered the story at length without quoting any Koko users, or citing complaints they may have posted on social media. When we asked Morris directly if Koko had received any complaints from users about the feature, or any questions or feedback regarding how the feature worked, he said it hadn't.

Ultimately all the X.com posters who were demanding lawsuits and calling for sanctions of some kind against Koko were doing so in response to purely speculative harms inflicted on imaginary users. It was a classic case of problemism at work.

In contrast, Rob Morris was focusing on what might be called the existential threat of the status quo. Broadly put, this refers to any shortcomings, harms, inefficiencies, or inequities that have existed for so long they've become fairly invisible. In this case, the specific threat was perennially unmet mental health care needs. In his efforts to develop

Koko, Morris is working to upend the status quo by broadening access to support and changing the provision of care.

This doesn't exempt him or Koko from scrutiny, criticism, and accountability. In fact, the opposite is arguably true. Asking what could possibly go right means committing to action, then iterating and learning from successes, failures, and criticisms alike. In addition, mental health care is a realm where questions around privacy, patient autonomy, and patient equity all require careful deliberation. Appropriate safeguards and transparent policies are necessary.

In this instance, though, Koko was acting transparently, in a carefully designed effort to explore AI-assisted peer support and improve accessible and effective care options for people. While critics loudly condemned hypothetical risks, Koko was committed to testing and refining evidence-based solutions.

The Doctor Will See You (Six Months from) Now

Rates for depression, anxiety, and suicide in the U.S. have risen dramatically over the last twenty-plus years, even as outcomes for conditions like cancer and heart disease are improving.[4] In 2022, nearly 50,000 Americans took their own lives, at the highest rate per capita since 1941.[5] An additional 100,000-plus[6] died from drug overdose deaths, which are strongly correlated with mental health issues.[7]

If statistics like these stand as the starkest indicators of what is now commonly regarded as an ongoing mental health crisis, they're hardly the only ones. Studies show that mental health issues increase mortality risks of all kinds, not just suicide, and can result in 10- to 20-year reductions in life expectancy.[8] Mental disorders have been correlated with lower GPAs and higher dropout rates among college students.[9]

Employees who rate their mental health as "fair or poor" report about "four times more unplanned absences due to poor mental health than do their counterparts who report good, very good or excellent mental health."[10] People who experience repeated job loss due to persistent mental health disorders can fall into a cycle of prolonged unemployment, as the loss of skills, professional networks, and confidence, along with employment gaps and potential employer biases, make it increasingly difficult to find and maintain new jobs. A 2011 report from the World Economic Forum projected that the annual direct and indirect global costs of poor mental health will reach $6 trillion in 2030.[11]

Reporting from the *Washington Post* shows that 129 million Americans live in places that qualify as professional shortage areas for mental health care. Fewer than one-third of the U.S. population lives in places that have sufficient access to mental health professionals to meet local needs. Consequently, it can often take as long as three to six months to obtain care from a qualified practitioner.[12]

Using technology and automation to broaden access to mental health care beyond the archetypal fifty-minute in-person appointment is a long-standing practice. Crisis hotlines have existed since the 1950s.[13] Teletherapy, online support groups, self-guided mental health apps, and online counseling platforms like BetterHelp and Talkspace have all become increasingly prevalent in recent years, especially during the Covid-19 pandemic. There are now more than 10,000 mental health apps, featuring self-guided therapy sessions, mindfulness and meditation exercises, mood tracking, CBT techniques, and more.[14]

And yet access to care has remained elusive for large swaths of the population. Important federal laws requiring health insurers to cover mental health services more broadly were passed in 2008 and 2010, but treatment gaps persist.[15] A 2018 report from the University of Michigan's School of Public Health found that 73 percent of the nation's 3,135 counties did not have a single child-and-adolescent psychiatrist.[16] A 2019 survey conducted by the Kaiser Family Foundation found that "half of all Californians [52%] say their community does not have enough mental health providers to meet its needs."[17] In October 2023, ABC News reported that call volume to 988, the National Suicide Prevention Hotline, had grown so heavy that some of the national network's call centers had begun to put a twenty-minute time limit on calls. Others only allowed individuals to call a maximum of three times in a given time period.

The existing lack of resources becomes even more pronounced when you consider that many people only seek out mental health services when experiencing significant challenges or distress. If we were to shift toward a more proactive approach, viewing mental health care as an ongoing component of overall wellness—a resource for times of relative well-being *and* crisis—the gap between available care and potential demand would likely be even greater than we generally perceive.

In addition, the mere existence of thousands of mental health smart apps is no guarantee of real-world impact. As with any kind of medical intervention, digital or otherwise, a key first step in achieving measurable improvements in patient outcomes involves clinical validation in randomized control trials. For example, researchers might test to see if users of a specific mental health app achieve lower anxiety scores on a standardized assessment like the Generalized Anxiety Disorder–7 scale after using the app for a given period of time.

But even accessibility plus proven efficacy means little if people don't use a product, or don't stick with it after they start using it. What really matters is what health care practitioners call uptake and engagement. And what many mental health care studies have shown over the last two decades is that the largely self-guided nature of apps leads to low levels of engagement and high attrition rates.[18] A 2019 study of such apps found that retention rates dropped to a median of 3.9 percent after just fifteen days of usage.

One way to improve engagement and retention rates is to incorporate more conversational or social elements into apps and other interventions. After all, one of the reasons traditional talk therapy is the mainstay of psychotherapy is that people find talking to other people compelling; we're social animals. That's what Morris is trying to do with Koko, through peer support. It's also what the developers of apps like Woebot and Wysa are doing via chatbots that approximate traditional talk therapy by engaging in ongoing dialogue with users.

While specialized mental health chatbots like Woebot and Wysa incorporate advanced natural language processing, they're not generative in the way that general-purpose foundation models like GPT-4 are. Instead, they rely on predefined structures called frames. Frames are what give traditional chatbots the somewhat rigid and predictable quality of a competent telemarketer working their way through a script, albeit one with multiple branching paths.

Thus, while these specialized mental health chatbots have some capacity to provide empathetic responses and simulate a therapeutic conversation, they're ultimately fairly limited in how they can respond to user input. On the one hand, this means they're not prone to hallucinations like ChatGPT and its peers—their source content is based on evidence-based therapeutic techniques vetted by mental health professionals. The downside, however, is that their capacity to respond in ways that are best suited for a given context or user declaration is not as nuanced or adaptable as human therapists or more advanced AI models.

"Whenever there is a place for unscripted, natural language input from the user," the authors of a June 2023 study in the journal *Front Psychiatry* write, "a chatbot has to classify it as falling into one of the coarse-grained, pre-defined categories."[19] Here's an example the authors cite to show how that can turn out when a user expresses a subtle but clinically significant indication of emotional distress that doesn't contain language the chatbot has been programmed to recognize:

1. Woebot: Okay, want to tell me more about what's going on?
2. User: I'm feeling weird. It's like I'm empty inside. When I look at the faces of my friends and family, I feel like they are empty as well. That's why we cannot connect with each other. There's nothing to hold on to.
3. Woebot: Sounds like you are dealing with feeling low . . .

In an instance that calls for an emotionally attuned response designed to leave a user feeling truly seen and heard, Woebot sounds like it is phoning it in. Now imagine if a user is in the middle of a panic attack or experiencing an impulse to harm himself. In such charged scenarios, you don't want a help-line volunteer with the emotional IQ of a hotel reservation chatbot. You don't want a mindfulness worksheet. And you definitely don't want to be told to call back in two weeks, when your new monthly call allotment kicks in.

What you want, instead, is an intelligent, emotionally responsive connection, available to you immediately, for however long you require it. For most people, another actual human may be the ideal help provider here. But what happens when no humans are available, or can't stick with you for as long as you feel you need support? Or what if you like the idea of a conversational partner who will never inadvertently gasp at a disclosure you make, or yawn at an inopportune time?

State-of-the-art LLMs aren't going to replace human mental health care providers anytime soon. In fact, as we'll discuss in the next section, there are great opportunities for human providers to use them in a wide range of ways to amplify their impact and increase the value of their work. But LLMs also create broad new possibilities—not just to replicate the skills and approaches of human providers, but also to reimagine how we can develop, test, and deploy mental health care as abundant intelligence becomes the norm. In this way, we might finally make real progress on the existential threat of the status quo.

Clinician, Know Thyself

An eye surgeon who replaces a cataract with an artificial lens can change a person's life in half an hour. In many health care domains, practitioners can assess the efficacy of their interventions in similarly tangible ways, using clear metrics like blood tests, bone scans, and other physiological markers.

In the domain of mental health care, the data informing diagnoses and the method of treatment often consists largely of words, incrementally accrued. While standardized questionnaires and rating scales provide some quantifiable measures, these still rely on self-reporting or clinician judgment. That leaves many opportunities for gaps in the narrative, misinterpretation, and cognitive distortions along the way. A patient maintaining a daily log of her moods, activities, or behaviors may not accurately report them, for any number of reasons. Clinicians who interact with patients in limited contexts, for relatively short periods of time, may draw incorrect conclusions about their conditions.

Digital technologies can help counteract these factors by providing more objective and continuous data collection methods. Now it's possible to passively monitor behavior through smartphones and wearable devices. Mental health apps that proactively prompt users about their moods and behaviors can help people achieve more consistent self-monitoring. AI systems that use geolocation data, texting frequency, and call durations can predict the onset of depression or bipolar episodes.

AI systems can also analyze vast collections of therapy session transcripts to better understand which kinds of interventions work best in different contexts, and which therapist behaviors lead to successful treatment. In January 2024, for example, researchers affiliated with Lyssn.io, a quality assurance and clinician training platform, and Talkspace, an online therapy provider, published a study in *JAMA Network Open* regarding their AI-driven analysis of more than 160,000 anonymized, text-based counseling sessions that had taken place on Talkspace between 2014 and 2019.[20]

In those counseling sessions, patients exchanged more than 20 million messages with licensed therapists. Using text-classifiers developed by Lyssn.io, the researchers were able to categorize all those messages on a very granular level, in terms of both the topics that were being discussed and the specific kinds of therapist behaviors that were being deployed. Any given message could be tagged with codes like "work," "parenting," "asking open-ended questions," and "making reflective listening statements."

The researchers could then link these coded utterances with data regarding patient symptoms, client satisfaction levels, the duration of client engagements, and symptom change as expressed through a client self-assessment questionnaire for depressive symptoms known as the PHQ-8. In this way, they could gain new insights about which kinds of therapist behaviors and interventions worked best in which specific contexts.

For example, this data revealed that while "giving information" was the most common therapist intervention, occurring in over half of therapist messages, it had a small but significant association with poorer outcomes. This finding helps illustrate the power of analyzing vast datasets, as smaller studies might not have surfaced the counterintuitive insight that providing too much psychoeducation can be detrimental to treatment progress.

In contrast, interventions like "complex reflections," or restatements that add substantial meaning or emphasis to what the client has said, and "affirmations," or statements that recognize and validate the client's strengths and efforts, were associated with better outcomes. By revealing patterns and relationships that would be unlikely in individual cases, or even dozens or hundreds of cases, such large-scale analyses can help practitioners better understand the subtleties of effective therapy and improve their therapeutic practices.

In the journal *npj Mental Health Research,* computational clinical psychologist Elizabeth C. Stade and her coauthors explore how mental health practitioners might start incorporating LLMs as assistants in their practices.[21] In a kind of specialized form of iterative deployment, they envision three potential stages of AI integration in the mental health space. Stage 1 involves relatively simple assistive uses of AI, such as using session recordings and a psychotherapist's own real-time notes to draft the more formal session notes, treatment plans, and other administrative documentation that psychotherapists must maintain for every patient they see.

Stage 2 involves more collaborative engagements, such as reviewing entire session transcripts to assess how well trainee psychotherapists are adhering to EBPs, or offering assistance and guidance to clients with between-session homework assignments, such as completing different kinds of CBT worksheets. Stage 3 involves fully autonomous care: clinical LLMs—trained on EBPs and rigorously evaluated for utility, efficacy, and safety—that can administer all the tasks and therapeutic interventions that human clinicians perform.

While some may anticipate job displacement as we move toward Stage 3, an equally plausible and far more ambitious narrative is possible too: one where mental health care is provided in the same way we provide music at Spotify and video at Netflix. Which is to say, accessibly, economically, scalably, informed by data analytics, and highly configurable to the unique preferences and needs of individual users.

A recurring theme in the internet age is that while Big Tech *could* use its considerable resources to tackle truly important societal issues, it prefers more trivial but lucrative challenges. "The best minds of my generation are thinking about how to make people click ads," Jeff Hammerbacher, an early Facebook employee, told *BusinessWeek* in 2011. "That sucks."

Here, however, is an opportunity to use data science to do more than just optimize cat-video discovery or accurately predict when demand for pink-flamingo pool floats will start surging in Miami Beach.

People choose mental health care practitioners for many different reasons, but it often boils down to convenience and pragmatism. Who's in your insurance network? Who did a trusted friend or family recommend? Who's in your area? Are they even taking new clients? Think about how that could change as abundant intelligence takes hold. When mental health care is truly affordable and accessible, therapeutic approaches and expertise can take precedence.

In many ways, existing mental health platforms like BetterHelp and Talkspace already pursue this approach. But their exclusive reliance on human practitioners imposes certain limitations. Service is not as on-demand as it is with Spotify or Netflix. You still need to schedule appointments. Even if you choose asynchronous text messaging as your preferred form of engagement, your therapist will likely have set hours in which they respond to replies; they're not working 24/7. The number of clients any one human therapist can manage is also limited.

Finally, if you don't have insurance, you will definitely end up paying significantly more than you do for Netflix Premium.

But a platform where there are wholly virtual therapists, along with human ones, would function in much less constrained ways. So how does care change when you can access a clinically tested and proven therapist whenever you want, for two minutes or two hours, for a flat monthly fee of $19.99? Wikipedia lists nearly 200 different kinds of psychotherapies. Other sources suggest there are more than 500.[22] Maybe analytical psychology is the best approach for you. Maybe dialectical behavior theory is. How do you know?

Since cost is no longer an issue, just start exploring. Very quickly you could test-drive multiple virtual therapists, and thus get a better sense of what you think will work best for you. If that means structured appointments, you choose those. If you want to be able to access your therapist in spontaneous, free-form ways, the way you look up things on Wikipedia, you do that. Maybe you regularly see more than one therapist. Maybe you assemble a team of therapists who you meet with jointly, so that you can benefit from second and third opinions in real time.

Searching for your perfect therapeutic fit would be easier than ever—and you might not actually have to do much searching. Data sharing in medical contexts is a sensitive and contentious issue, but it can also significantly improve treatment personalization. An AI-powered mental health platform could analyze data from thousands of patients who share your age range, gender, and specific anxiety symptoms.

A platform like this would know which cognitive behavioral therapy exercises or medication regimes have led to the most significant improvements for others similar to you. It could prioritize approaches that have surpassed certain thresholds of efficacy in real-world scenarios, and recommend personalized treatment plans. It could even present you with some different choices, share their respective success rates with you, and allow you to pick the one that most appeals to you. Maybe every practitioner on the platform, human and AI alike, would have user reviews appended to their profile, like the ones on Amazon. Eventually, the platform might offer a Spotify-inspired "Therapy Mix" that curates a unique blend of different therapeutic approaches based on your engagement history.

Would all this trivialize psychotherapeutic care? Only if we think that it's better for society if mental health care is more difficult to access, less informed by data analytics, and less personalized than Netflix content. Could it introduce new ethical challenges? It might, and that's why transparency, iterative deployment, and rigorous evaluation of any deployment efforts are so necessary in this domain.

Even if we manage all that, though, there's still a case to be made that the therapeutic alliance that exists between patients and their therapists—the sense of empathy and trust that develops over time, through ongoing interaction—is something that we shouldn't automate completely, no matter how reliable and discreet and clinically validated therapy LLMs may become. Some might argue that, in choosing this path, we would lose some fundamental part of our humanity.

But this is problemist thinking, focused on only one side of the ledger. Let's consider the other side: AI systems that can help train more human practitioners than we're capable of training now. AI systems that can support human practitioners in ways that will allow them to interact with more patients than they currently can manage. AI systems that will enable millions of people who are underserved by current approaches to obtain abundant and affordable care. Or do we think it's more appropriate to continue to disadvantage the underserved in pursuit of retaining our humanity?

Another problemist trope is the idea that care that becomes *too accessible* and *too affordable* could lead to an overreliance on therapeutic LLMs and a reduction in individual agency. But this concern is largely based on how we define the scope and utility of mental health care—or, more specifically, on how we've traditionally defined it. In part because human-provided care is costly to supply, we see it primarily as a reactive mechanism for responding to acute problems. If that's the baseline, then it's easy to interpret more frequent usage as overreliance. But how often do you hear people raising concerns about eyeglasses addiction, or a problematic dependence on pacemakers or seat belts?

What if we look at the prospect of proactive mental health care and AI models providing emotional support on tap? When we shift to a what-could-possibly-go-right mindset, the benefits start stacking up. People feel seen when they most need to be seen. They gain access to empathetic responses and helpful strategies at any time, day or night. They develop healthier coping mechanisms and emotional resilience through consistent, personalized support.

This ambitious vision doesn't merely aim to replicate and scale up current therapeutic practices. Instead, it aims to transform and enhance them, potentially ushering in a new era of mental health care that is more comprehensive, continuous, and deeply integrated into daily life.

Superhumane

When new innovations are so powerful that they start reshaping social norms and individual behaviors, Gloomers often cast such transformations as inherently destructive. Before most people even knew what "WWW" stood for, there were books about "internet addiction." Why? Because the idea of millions of people suddenly spending hours typing to anonymous strangers didn't fit into existing conventions about healthy social interaction and productive time management.

As people begin to engage more frequently and meaningfully with LLMs of all kinds, including therapeutic ones, it's worth noting that one of our most enduring human behaviors involves forming incredibly close and important bonds with nonhuman intelligences. Billions of people say they have a personal relationship with God or other religious deities, most of whom are envisioned as superintelligences whose powers of perception and habits of mind are not fully discernible to us mortals. Billions of people forge some of their most meaningful relationships with dogs, cats, and other animals that have a relatively limited range of communicative powers. Children do this with dolls, stuffed animals, and imaginary friends. That we might be quick to develop deep and lasting bonds with intelligences that are just as expressive and responsive as we are seems inevitable, a sign of human nature more than technological overreach.

In most instances, we consider our capacity to cultivate relationships with nonhuman entities as one of our most valuable attributes, because of how these relationships can increase our emotional intelligence and complement and impact our relationships with other people. Many of these relationships function as a source of emotional support without judgment. They help create contexts where people feel comfortable enough to express themselves candidly. They often provide a sense of purpose, contributing to overall well-being.

Throughout 2023, as millions of people interacted with LLMs for the first time, an ongoing refrain in the discourse, from critics, advocates, and the models themselves, if you asked them, was that models have no real understanding of the world, and thus no emotional intelligence and no real insight into or comprehension of the feelings or states of mind of those with whom they were interacting. No empathy, in short. And that this was one reason it was so important to "keep humans in the loop."

And yet very quickly many people were discovering how responsive, present, patient, and seemingly emotionally attuned conservational AI models can be. Consider a study from April 2023 published in *JAMA Internal Medicine.*[23] The study's authors randomly selected 195 conversations from a Reddit subreddit called AskDocs, where Reddit users ask questions and verified physicians answer them. It asked ChatGPT to answer them as well, then asked panels of three licensed physicians (not psychotherapists) to rate the two different answers for each of the 195 conversations—one from ChatGPT and one from the human caregiver—without identifying who wrote what. In 78.6 percent of the instances, the physicians rated ChatGPT's responses higher than the human responses.

In effect, while AI models are explicitly not conscious or self-aware, they are, in their own statistically probable way, performatively kind and empathetic in ways that often surpass human norms. The potential consequence of this was brought home for me in conversations I had on two different episodes of my *Possible* podcast: the first with my Inflection AI cofounder Mustafa Suleyman, and the second with Maja Mataric, a computer science professor at USC Viterbi School of Engineering, who designs socially assistive robots. Both stressed how different kinds of AI simulating empathy can end up having a profound impact on humanity at large. As Mustafa suggested, not everyone has reliable access to human kindness and support. But when that becomes something you do have "always on tap," it ends up increasing your own capacity for "being able to be kind to other people."

According to Maja, she and her colleagues have designed robots that "behave empathetically to the user and, in that way, make the user more empathetic." And as Maja explained, numerous studies have shown that behaving empathetically improves physical health.

So imagine the compounding effects we are likely to see over time as people integrate AI models more fully into their lives. Along with helping us unlock the secrets of protein folding, or increase the efficiency of renewable energy systems, what if our constant interactions with AI models endowed with the preternatural equanimity of the Dalai Lama simply help us become nicer, more patient, and more emotionally generous versions of ourselves? Through complex neural networks and massive global server clusters, the world could become superhumane.

CHAPTER 4

THE TRIUMPH OF THE PRIVATE COMMONS

Sure, superhumane mental health support that's always on tap may *sound* great—but what's the catch? Because there has to be a catch, right? Even if we give users clear and truly effective control over how their personal data is safeguarded and shared. Even if we strike the right balance between productively managing mental health challenges and resisting the urge to completely smooth away the everyday emotional tumult of a full human existence.

Even if we get everything right, won't Big Tech still somehow keep the bulk of the rewards to itself, at the expense of other stakeholders? This is, after all, the conventional wisdom on how Big Tech innovation typically plays out.

See, for example, *MIT Technology Review*'s take on this subject:

> For all the amazing advances in AI and other digital tools over the last decade, their record in improving prosperity and spurring widespread economic growth is discouraging. Although a few investors and entrepreneurs have become very rich, most

people haven't benefited. Some have even been automated out of their jobs.[1]

Here's the gifted science fiction writer Ted Chiang expressing a similar perspective in *The New Yorker*:

> In the United States, per-capita G.D.P. has almost doubled since 1980, while the median household income has lagged far behind. That period covers the information-technology revolution. This means that the economic value created by the personal computer and the Internet has mostly served to increase the wealth of the top one percent of the top one percent, instead of raising the standard of living for U.S. citizens as a whole.[2]

In the realm of Big Tech critique, especially where AI is involved, these assessments are fairly mild. For a more scathing point of view, consider the work of Shoshana Zuboff, professor emerita of the Harvard School of Business. In *The Age of Surveillance Capitalism*, an international bestseller widely heralded as one of the best books of 2019, Zuboff picks up where George Orwell left off in 1949. In her estimation, companies like Google and Facebook have replaced the Party and Big Brother with a "sensate, networked, computational infrastructure" that Zuboff brands "Big Other." Instead of technology weaponizing the state into a "project of total possession," it weaponized the market into a "project of total certainty."[3]

As Zuboff presents it, Big Other operationalizes ubiquitous surveillance to drain us of individual agency one geocode request at a time. Algorithms gorging on our personal data progressively undermine our freedom of will. Democracy optimizes—that is, implodes—into market-driven totalitarianism. In time, our misbegotten desire for traffic-free travel routes and positively Yelped pizza delivery vaporizes our capacity to live self-determining lives.

How did we arrive at this hell?

Early in its existence, Google realized that every action that users took on its site—the search strings they entered, the links they clicked on, and more—were trackable. All of this stuff, in some ways so seemingly devoid of value that it is sometimes described as "data exhaust" rather than data, could now be saved, aggregated, analyzed, recombined, and ultimately deployed in new ways, on a sweeping scale.

In this way, Google could improve user experience. If people searching for "best mp3 player" were consistently clicking on links to "iPod reviews" in their search results, Google could refine its ranking algorithms to prioritize product comparisons for portable music players in future searches related to digital audio devices. As Google learned that users typing in "wea" were probably intending to search for "weather forecast" or "weather today," it could suggest an auto-completion to save them a few keystrokes.

As long as users derived *all* the value from Google's efforts to productively leverage this behavioral data, Zuboff maintains, it was a fair exchange. In fact, in those early Edenic days, Zuboff suggests the company's approach was doubly virtuous. Along with using data to improve search, Google also used it to create entirely new products and services, like translation software. Zuboff calls this process the "behavioral value reinvestment cycle."[4]

In a quest to generate enough revenue to sustain its operations, however, Google eventually committed what Zuboff identifies as the original sin of "surveillance capitalism": it began to apply some of the behavioral data it was collecting to make ads more relevant to users. In doing so, Google's theory went, users would click on ads more often and advertisers derive more value from the process too.

"Operationally, this meant that Google would turn its own growing cache of behavioral data and its computational power and expertise toward the single task of matching ads with queries," Zuboff writes.[5] A few paragraphs later, she tempers the absolutism of this declaration by allowing that "*some* data would continue to be applied to service improvement, but the growing stores of collateral signals would be repurposed to improve the profitability of ads for both Google and its advertisers" (emphasis ours).[6]

Never mind that making ads more relevant to users and advertisers is itself a form of service improvement. To see where Zuboff's characterizations of Google's alleged intents and priorities really start diverging with reality, simply note that Gmail, Chrome, Maps, Street View, and Docs are all products that launched *after* Google implemented the ad-tech innovation that Zuboff positions as the moment the company broke faith with the "behavioral value reinvestment cycle."

To make her thesis plausible, Zuboff is compelled to maintain that the primary purpose of all these products is to generate "behavioral surplus"—that is, more data Google can use to weaponize its ad business to the point where targeting crosses over into behavior manipulation. Armed with these surpluses, Zuboff contends, Google can effectively operate "behavior prediction markets" where a person's future actions, forecasted with total certainty through algorithmic modeling, are sold to the highest bidder.

So, all those times you thought you were just confirming pickup times with your child's preschool or getting on-the-fly directions to that new coffee shop downtown? In Zuboff's telling, something far more sinister is afoot:

> Although Big Other can mimic intimacy through the tireless devotion of the One Voice–Amazon-Alexa's chirpy service, Google Assistant's reminders and endless information—do not mistake the soothing sounds for anything other than the exploitation of your needs. I think of elephants, the most majestic of all mammals: Big Other poaches our behavior for surplus and leaves behind all the meaning lodged in our bodies, our brains, and our beating hearts, not unlike the monstrous slaughter of elephants for ivory. Forget the cliché that if it's free, "You are the product." You are not the product; you are the abandoned carcass. The "product" derives from the surplus that is ripped from your life.[7]

That so many of us keep dragging our abandoned carcasses back for more, dozens of times a day, for years on end, suggests that Zuboff may be overstating her case when she claims that Big Tech companies "do not establish constructive producer-consumer reciprocities." Six of Google's products have more than two billion users each. About 1.46 billion people have iPhones.

Numbers like that send a strong signal that the value that Big Tech creates flows two ways—but Zuboff is adamant that that's not the case. Over the course of *The Age of Surveillance Capitalism,* she characterizes Big Tech and AI as "extraction operations" so frequently that the simple act of booking a meeting via Calendar begins to acquire the menace of a mountaintop blow-off or a fracking earthquake. Every time you use Maps to locate a nearby ATM stands as a brutal zero-sum plundering committed by Big Other.

Granted, computer scientists have themselves been using the word "extraction" to describe aspects of database retrieval since at least the

1950s. And apparently Zuboff herself extracted this terminology from essays written by Google's Hal Varian. But this usage has always been more metaphor than sound information theory. You simply don't "extract" data in the same way you extract oil, copper, or even a tooth. When you're blasting out tons of bituminous coal from a deep seam buried thousands of feet beneath the earth's surface, you irreversibly deplete that finite resource, and leave a hole in the ground to boot.

With digital files, only copies are taken. The originals remain intact and unchanged in their original location. We're also adding to our global data reserves much faster than it takes a decomposing *T. rex* to morph into a half tank of Chevron Supreme. Every hour now, we augment our already-undepletable reserves of selfies, Reddit posts, Facebook likes, marketing decks, Google searches, route planning, *Call of Duty* fanfics, medical studies, and YouTube videos with enough new content to fill a virtual supertanker.

At the same time, it's also true that the way many AI datasets are created is a highly contentious issue. To create a state-of-the-art LLM takes massive amounts of training data. OpenAI's GPT-3 was trained for 300 billion tokens.[8] GPT-4's training dataset was even larger, though OpenAI has not released specific details of its size.

Typically, developers start assembling their training data by relying on existing open repositories that aggregate vast amounts of data from crawling the web. One well-known dataset, called Common Crawl, which is maintained by a nonprofit of the same name, contains more than 2.7 billion web pages.[9] Another, known as the Pile, starts with a modified version of the Common Crawl dataset, then adds twenty-one additional subdatasets that incorporate material from Microsoft's coding website GitHub, scientific papers from PubMed Central and arXiv, various books and literature datasets, legal corpora from the Free Law Project, U.S. Patent Office backgrounds, YouTube subtitles, and more.[10]

Google has also created its own dataset, C4, which stands for Colossal Clean Crawled Corpus. A *Washington Post* analysis of its contents determined that its top five sources were patents.google.com, Wikipedia, the document-hosting site scribd.com, the *New York Times*' website, and PLOS, a nonprofit open-access publisher of scientific and medical research papers.[11]

As you might intuit from the enormous scale of these datasets generally, and from some of the specific sources that underlie them, all incorporate material from websites, books, and scientific publications without obtaining explicit consent from copyright holders of this material. Such usage has led to accusations of widespread copyright infringement.

In response, AI developers generally hold that their uses of data are both legal under existing copyright law and broadly beneficial to their users and society at large. To date, a number of plaintiffs, including the *New York Times,* Getty Images, and various authors and artists, have filed lawsuits alleging copyright infringements against AI developers like OpenAI, Microsoft, Stability AI, and Midjourney.

How these lawsuits will resolve is not yet clear. If courts determine that training on data to extract patterns and information, rather than to reproduce or incorporate an original work in recognizable forms, doesn't fall under fair use, we'll need novel solutions to manage licensing at such enormous scale. Given that virtually all content on the internet is automatically copyrighted, new mechanisms for clearing billions of blog posts, user comments, product reviews, photographs, and memes along with news articles, books, or feature movies will be needed. Such mechanisms would need to balance the interests of content creators, AI developers, and the public good.

In the meantime, what the internet era has shown us so far is that making broad and creative use of data generally creates tremendous value, for individual users and society at large, as well as for developers. When data that is dormant, underutilized, or only relevant in narrow contexts is repurposed, synthesized, and transformed in novel and compounding ways, that's not extractive. It's resourceful and regenerative.

So rather than "extraction operations," we see something more akin to data agriculture. Instead of Big Other usurping value from users, we see a mutualistic ecosystem of developers, platforms, users, and content creators whose interactions and contributions collectively enrich the lives of billions of people every day. In a sense, what we've seen in the commercial internet era is a new kind of *private commons*. In the age of AI, it's on the verge of growing even more fruitful.

Doublethink Different

What do you think of when you hear the term "commons," or its even more emphatic variant, the "public commons"? A picturesque town square in some quaint New England village? An urban garden in downtown Los Angeles whose collective owners all share equal access rights? Google Maps?

Probably not the latter. The most robust contemporary definitions of the "commons" generally refer to resources that are characterized by both shared open access and communal stewardship, for individual benefit and the benefit of a community as a whole. "The commons are property we all share, property that's owned not by any one person or group, but that's held, well, in common," writes Steven Lubar, a professor of American studies at Brown University, in *Smithsonian Magazine*.[12]

So the phrase "private commons" may hit your ear with an oxymoronic—or even Orwellian—lilt. But it's also true that on a conceptual level the commons has always had somewhat amorphous boundaries. In its earliest uses, "commons" referred to natural resources like pastures and forests, which local inhabitants could use for cattle-grazing, hunting, and other agrarian pursuits. Today, "commons" are often defined more broadly than that, especially in general public discourse. Public parks and beaches are frequently tagged as such, as are air, water, and public libraries. Creative works in the public domain are part of the public commons too, as are language itself, written alphabets, many computer languages, the recipe for an Old Fashioned, or even the view of Orion on a clear, dark night.

Among academics, "commons" are often more narrowly defined than in popular understanding. Elinor Ostrom, an influential political scientist who received the Nobel Prize in Economics in 2009 for her work on commons, identified eight principles that characterize successful "common-pool resource" institutions. In total, these principles create the blueprint for a communal resource that is significantly more circumscribed than air or the recipe for an Old Fashioned is. In Ostrom's conception, a commons is an intentionally managed resource with a defined community of users. This version of a commons has clearly defined boundaries of access, graduated sanctions for rule violations, and other explicitly articulated and operationalized governance features. In many ways, a commons is governed by the sort of rules we'd expect from a homeowners association.

Broader conceptions of common resources leave more room for variance and often place more emphasis on open access than governance. Sometimes a resource is owned by all (or none), such as the recipe for an Old Fashioned or that view of Orion. In other instances, the resource has an owner, or collective ownership, but it still remains broadly accessible. For example, public parks and public libraries might be owned by local governments and funded by local taxpayers. They might charge usage fees of various sorts but they remain accessible to anyone, at least to some extent. You can't check out a book from the New York Public Library if you're not a certified resident of New York City with a valid library card. But you can still enter it freely and pass the day on its premises reading its copy of Milton Friedman's *Capitalism and Freedom* without spending a penny.

So where might you place platforms like Google Maps or Yelp in *this* commons cosmology? In many respects, they function very much like Wikipedia does, with volunteer users contributing information that enriches these platforms for all who use them. All are freely accessible to very large audiences. All operate under specific governance rules and are explicitly managed by paid staffs. But Google Maps and Yelp, by virtue of their profit-driven nature, are understandably classified as proprietary commercial services. Wikipedia, in turn, is often accorded commons status because the organization that develops and manages it is a nonprofit foundation.

To account for them all in a way that reflects their essential commonalities, you could describe them all as private commons. Since the internet's first commercial stirrings in the early 1990s, privately owned or administrated platforms that enlist users as producers and stewards have proliferated. Various labels have been applied to different aspects and instances of this template, including Web 2.0, social media, the sharing economy, the gig economy, and surveillance capitalism. But none of these—approving or pejorative—aptly convey the emergence of free and near-free life-management resources that effectively function as privatized social services and utilities, the welfare state moving at the speed of capitalism. The term *private commons* does this.

Every time you search for something using Google, or add a calendar event to Calendar, or obtain driving directions and traffic reports from Waze, or look for apartments on Craigslist, or store photos on Dropbox, you're benefiting from the private commons. The same is true when you solicit career advice from professionals in your industry on LinkedIn, or watch a YouTube video that shows you how to fix your leaky kitchen sink, or listen to a podcast on Spotify Free, or punctuate an email with an assortment of emojis.

While for-profit corporations and other private institutions play a vital role in the creation of the private commons, the public obviously does too. On Facebook, on YouTube, on X.com and all of the others, individual users provide much of the content, all of the eyeballs, and all of the user behavior, in the form of clicks, social interactions, purchases, and more, that help platform operators turn a profit.

This symbiotic relationship between users and platforms creates unprecedented value. But it also creates ongoing questions about how this value is distributed. When people see Big Tech companies accruing billions of dollars a year in ad revenues, they often want to know where their cut of the action is. And understandably so. In 2023, Alphabet, aka Google, reported a profit of $73.7 billion. Microsoft's was $72.4 billion. In *Forbes*'s "World's Billionaire List" for 2024, seven of the ten individuals denoted as "the richest people in the world" generated their fortunes via tech.[13]

For users, the value they accrue from the private commons lacks the explicit visibility of dollars and cents, and this is at least one of the reasons the perception that Big Tech withholds most of the value it creates for itself is as widely held as it is. But that doesn't mean a more reciprocal value exchange isn't happening.

Economists call the difference between what people pay for a product or service and how much they value it "consumer surplus." If you buy a jacket for $100, but think it's worth $200, the consumer surplus in that case is $100.

When a product or a service is provided for free, a consumer surplus can exist too, as long as the consumer places some value on the offering. In this way, broadcast TV and radio have been major sources of consumer surplus for decades. You pay nothing for them beyond the cost of your TV set or radio, and get a lifetime of entertainment and information.

Recognizing that the digital economy is difficult to measure using traditional means because it offers so many of its products and services for free, Erik Brynjolfsson, the director of the Stanford Digital Economy Lab, and Avinash Collis, an associate professor at Carnegie Mellon University, devised a series of "massive online choice experiments" designed to measure consumer surplus in the internet realm. Here's how they described these experiments in a 2019 article they wrote for the *Harvard Business Review*:

> To measure the consumer surplus generated by Facebook, we recruited a representative sample of the platform's U.S.-based users and offered them varying amounts of money to give it up for a month. To verify the responses, some participants were randomly selected to actually receive payments and forgo the service for the month. We temporarily added them as Facebook friends—with their permission, of course—to confirm that they didn't log in for that month.
>
> Some 20% of the users agreed to stop using the service for as little as $1; an equal proportion refused to give it up for less than $1,000. The median compensation our Facebook users were willing to accept to give up the service for one month was $48.[14]

What Brynjolfsson and Collis found was that the internet is basically a consumer surplus-generating machine. Along with running experiments focused on specific sites, like Facebook and Wikipedia, they also made offers to users for broader categories too, including email, search engines, maps, e-commerce, video, music, social media, and instant messaging.

According to their survey data, the median amount that it would take people to give up using search engines for a year was a whopping $17,530. For email, it was $8,414. For digital maps, it was $3,648.[15]

As noteworthy as these high appraisals are, there's a strong case that even they don't fully reflect the value the private commons brings to our lives. An observation Brynjolfsson and Collis make in their *HBR* article points to why: "*Britannica* used to cost several thousand dollars, meaning its customers considered it to be worth at least that amount," they write. "Wikipedia, a free service, has far more articles, at comparable quality, than *Britannica* ever did."

In other words, it's not just that Wikipedia replaces what was once a fairly expensive product for free. It's a *better* product too, because it contains significantly more articles. This makes it more useful to any individual user, and it also makes it more relevant to a wider number of users. How much more useful? In 1990, its best sales year ever, *Encyclopaedia Britannica* sold 120,000 copies in the U.S. In comparison, Wikipedia records around 4 billion monthly page-views in the U.S. alone.

Along with being incredibly useful, Wikipedia is also incredibly easy to use. Even if you were history's most devoted *Britannica* user, with a complete set in your home study, another in your office at work, and copies of the condensed one-volume version in your bathroom and your car, your ability to consult it whenever you wanted would still not compare to how easy it is to access Wikipedia today.

Digital Free-for-All

The more we apply hard numbers to the value the private commons generates, the more we begin to see how much that value compounds through this remarkable resource. Consider this excerpt from an interview that the American Enterprise Institute did with Hal Varian in 2017:

> VARIAN: Let's take an example of photography, all right? So in 2000, there were 80 billion photos produced. We know that because there were only three companies that produced film. And fast-forward to 2015, there are about 1.6 trillion photos produced. Back in 2000, photos cost about 50 cents apiece. Now they cost zero apiece essentially. So any ordinary person would say, wow, what a fantastic increase in productivity, because we've got a huge amount of more output and we've got a much, much lower cost.[16]

In this instance, the economic benefit for 2015's photos alone is $800 billion—that is, 1.6 trillion photos x 50 cents per photo to develop.

Now, it's no doubt true that many, and maybe even most, of the digital photos that were taken in 2015 were not ones anyone would have even bothered to take if they'd had to pay for them. Prior to the iPhone era, how many people ever used their Polaroid Instamatics to remember the spot where they left their car in a parking garage? But that's part of what makes this particular innovation so valuable. It's not just that digital photography reduces the costs of things people used to pay for, essentially to nothing. Even more profound is the fact that these new technologies allow us to do entirely new things, like using photographs for note-taking.

It's also not just the photographer saving fifty cents per snap who accrues value from this new capability. Because billions of people now have digital cameras in their phones, and the cost of each image we make is effectively zero, we collectively take trillions of photos a year.

We also now share a huge number of these images with other people, including complete strangers, because the private commons provides storage and distribution services for free too.

As much value as that creates for us as photographers, it arguably creates even more value for us as consumers of information. Now, if you want a peek in advance at the interior of some hotel in Istanbul where you're thinking about staying, candid third-party images of it are probably accessible somewhere. If you want to know who went to that party you couldn't attend, you can find that out.

What is the value of this completely unprecedented access to global Big Knowledge that billions of people now take for granted?

While it's impossible to calculate, it manifests in endlessly evolving and highly useful ways. In the early years of YouTube, critics were quick to characterize it as infinite "short-attention-span fluff."[17] Then, in just one of its many surprising metamorphoses, YouTube started functioning as a visually driven, applied-knowledge Wikipedia. If, in the wake of some supervolcanic eruption that left only you and YouTube standing, you could use its boundless catalog of how-to videos to rebuild civilization from scratch, one barndominium and three-ingredient dinner at a time. Once dismissed as yet another sign of humanity's long slide into inanity, YouTube now stands as a virtual storehouse of human knowledge so comprehensive it makes the Library of Congress look like a box of old John Grisham novels left out on trash day.

Of course, a little bit of knowledge can be a dangerous thing, right? Echo chambers, filter bubbles, and algorithmic radicalization red-pilling impressionable young gamers into increasingly extreme and destructive viewpoints is certainly a narrative that has gotten significant media coverage. But there's also a less covered story, even though it works exactly the same way: algorithmic springboarding. That's what happens when YouTube's recommendation algorithms lead users down spiraling rabbit holes of education, self-improvement, and career advancement.

Let's say you're a recent high school graduate trying to figure out your next move. You spend a lot of time playing *Tekken* and *Final Fantasy*, trading jokes on Reddit forums, and aimlessly surfing the internet—so much so that eventually you become interested in the technologies underlying it. Intrigued, you watch a "Python for Beginners" video hosted by some highly entertaining full-stack guru and find yourself hooked. One video leads to another. Because YouTube prioritizes watch-time and often pushes in-depth, lengthy videos, you're led down a bread-crumb trail of insidiously instructive programming tutorials. It's the same algorithmic logic that's been accused of reinforcing extremist beliefs, yet here it's forging a path to a better future.

Eventually, you head to Python.org and download the Python interpreter, which is free. Eager to experiment, you then seek additional help on GitHub, where you find repositories with beginner-friendly projects and engage with the coding community for tips and critiques. You also discover platforms like Stack Overflow and freeCodeCamp, offering forums for troubleshooting and free courses, respectively. These resources help you write your first Python script—a data visualization tool that creates simple charts from CSV files, inspired by your interest in tracking in-game statistics from your favorite video games.

In this journey, each element of the private commons—from YouTube tutorials to open-source communities on GitHub, to free courses on freeCodeCamp, to professional networking on LinkedIn—contributes synergistically, creating a tailored pathway that transforms your initial curiosity into employable skills.

And your story doesn't need to end there. At some point you apply for an entry-level data analysis position and present your dataviz tool as a part of your portfolio. Your initiative pays off, and you land the job.

As you gain experience, you create a LinkedIn profile highlighting your programming expertise, data-analysis skills, and your knack for automating processes. In no time, a startup spots your profile and sees the value you can bring. You receive an offer for a more challenging and rewarding role, and thus take another step closer to defining your place in the world.

Democratizing access to knowledge and opportunities, the private commons enables individual agency, educational opportunity, social mobility, and, ultimately, professional growth. And in at least one respect, its upside is limitless. That's because digital commons function very differently from the shared grazing pastures and other natural resources that comprised the oldest forms of traditional commons.

In the case of a sheep meadow or a fishing ground, outcomes can be subject to a perverse dynamic, dubbed "the tragedy of the commons" by the ecologist Garrett Hardin in a 1968 essay. The more useful or valuable people find a shared resource, Hardin reasoned, the more likely they will collectively ruin it through overuse. That's not how it works with digital resources, however, no matter how many times Gloomers describe Big Tech platforms as "extraction operations."

Data Wants to Be Useful

Garrett Hardin had deep concerns about "extraction operations" himself—the old-fashioned kind involving actual natural resources. A professor of ecology at the University of California, Santa Barbara, Hardin wrote his essay "The Tragedy of the Commons" during a moment when, even by contemporary standards, projections of imminent eco-doom were especially apocalyptic.

In *The Population Bomb*, also published in 1968, Stanford professor Paul Ehrlich famously warned that "hundreds of millions of people" would starve to death in the 1970s regardless of "any crash programs" deployed to prevent this disaster.[18] Hardin's take was no less dire. "An implicit and almost universal assumption of discussions published in professional and semipopular scientific journals is that the problem under discussion has a technical solution," he wrote in *Science*. But in his estimation, this simply wasn't true for the problem of overpopulation in a world of finite resources. "[People] think that farming the seas or developing new strains of wheat will solve the problem—technologically. I try to show here that the solution they seek cannot be found."[19]

Hardin went on to assert that individuals always try to maximize their own gain when taking advantage of a common resource. This is the "tragedy" of the commons: a herdsman with access to a communal open pasture will always add one more head of cattle to his herd if he can do so. It's the rational choice. Any benefit he derives from that extra animal—such as selling it at auction—will go entirely to him. Meanwhile, the cost of his decision—a slight reduction in the amount of grass available for communal grazing—will be borne by every herdsman who makes use of that shared resource. With every herdsman behaving in this same rational manner, Hardin reasoned, commons face inevitable depletion. "Each man is locked into a system that compels him to increase his herd without limit—in a world that is limited," Hardin concluded. "Ruin is the destination toward which all men rush, each pursuing his own best interest in a society that believes in the freedom of the commons. Freedom in a commons brings ruin to all."*

* As it turns out, the food production challenges that prompted Hardin to write his essay were significantly mitigated by technological interventions—including the development of higher-yield crops, improved irrigation techniques, and the increased use of fertilizers and pesticides. While food insecurity remains a serious problem in many parts of the world, global population is now more than twice what it was in 1968. Imminent mass starvation due to overpopulation, which Hardin and many others were predicting in that era, did not come to pass.

Hardin believed that the only ways to solve this dilemma included "private property, or something formally like it," or "coercive laws or taxing devices." Given his fundamental anti-tech, anti-growth stance, his work has ironically served as a favored touchpoint of real-estate developers looking to turn patchy, wind-eroded pastures into verdant golf courses. Meanwhile, his contention that any open-access commons, be it national park or free-range buffalo herd, is inexorably headed toward collapse also helped renew interest in traditional commons approaches to resource management. Elinor Ostrom, in particular, spent years demonstrating through extensive field research that local communities could, and often did, effectively manage common-pool resources sustainably without resorting to privatization or government oversight.

In its early years, the internet itself existed as a new kind of common-pool resource. Sure, someone owned the servers, routers, switches, and fiber-optic cables that turned code into virtual terrain you could freely explore and make use of in various ways. Technological limitations also served as a kind of inherent regulation; there wasn't that much you could do there. But there was also very little formal oversight. Within the boundaries of what communications protocols like TCP/IP and HTTP made possible, you pretty much had the latitude to do whatever you wanted.

As the technologies evolved, a wider range of actions became possible. Commercial interests began to sense how they could profitably develop this wide-open frontier. Government regulators realized that people had begun to build an entirely new world without them. As these new interlopers began to play a greater role in this domain, the desire for digital commons—online spaces explicitly resistant to commercial imperatives and government oversight—became even more pronounced. The open-source software movement gained momentum and became more relevant than ever. A nonprofit called Creative Commons developed a set of licenses that enable content creators to share their work more flexibly than traditional copyright permits.

With its emphasis on peer-based governance, the commons approach articulated by scholars like Ostrom is in many ways well suited for the internet. And yet, the heavy emphasis it places on the allocation of scarce resources in the face of intense and enduring demand is also a misleading framework for conceptualizing the challenges and opportunities of the digital world.

That's because platforms like Google Maps or Waze do not face the same challenges as a pasture surrounded by herdsmen looking for the cheapest way to feed their cows. Unlike a grazing field where each cow's consumption leaves less grass for other cattle, you can use digital platforms extensively without diminishing others' ability to use them. In fact, the reverse is true. If one thousand people use Waze in your city to navigate and report traffic conditions, it's a useful tool. If ten thousand people use it, it's even better. So, a digital commons can function very differently than traditional physical commons. Instead of carefully controlling access to scarce and hard-to-replace resources like timber or salmon, an obvious strategy is for digital commons to treat the resources that generate their value—data—as shared nonrivalrous resources that should be cultivated as much as possible and used proactively.

This paves the way for private commons, and especially private commons administered by large, well-provisioned organizations. Much of the value of private commons comes from the users, but it's the companies that provide the platforms and technology to run these digital common spaces that have the means to bring so many users together.

And yet obviously there are different opinions regarding the best way to conceive of the data that users generate in such contexts. The basic contract that powers much of the private commons today is that users get free services in return for giving platform operators access to the data that they generate. But the massive revenues that the biggest Big Tech companies amass, coupled with ongoing narratives around the value of privacy, most explicitly concretized in legislation like the EU's General Data Protection Regulation and the California Consumer Privacy Act, provoke questions of fairness.* Is free access to products and services a genuinely equitable exchange?

According to Statista, Meta's annual revenue per user (ARPU) in 2023 was $44.60.[20] This metric varies across countries and Meta's different platforms (Facebook, Messenger, Instagram, and WhatsApp). An Instagram user in France might generate more revenue for Meta than a Facebook user in Japan, or vice versa. On average, though, they're each worth approximately $44.60 a year, or $3.71 per month, to Meta. We also know from the research of Erik Brynjolfsson and Avinash Collis that the median compensation that Facebook users were willing to accept to give up the service for one month was $48.

Those numbers suggest a classic win-win proposition. On average, the value that users feel they're getting is more than twelve times the value they create for Meta. But the exchange still gives Meta ample resources to maintain the services it provides, invest substantially in the development of new ones (including many open-source AI models it makes available to the world for free), and generate a return for its shareholders.

Of course, not all users generate the same exact amount of value

* Implemented by the European Union in 2018, the General Data Protection Regulation aims to give individuals control over their personal data and harmonize data privacy laws across Europe. The California Consumer Privacy Act, enacted in 2020, is similar. Both regulations require companies to be more transparent about data collection and usage, and give users more control over their personal information.

for a platform. Extremely engaged users may create more posts, likes, reviews, ratings, and other kinds of data than less active users. Some users create blog posts, videos, and other kinds of content that drive high levels of engagement from other users.

Similarly, not all data is equally valuable, nor do users invest equal amounts of time and effort in creating it. Data that you generate passively when you log on, stay on a site for a given amount of time, and engage in various activities there may not require much effort on your part. Comments you dash off in reply to someone else's Instagram post may represent a greater investment of time and thought, but are still not the same as the fifteen hours you might invest in producing a well-crafted YouTube video.

It's also true that context matters. Your geolocation data might have minimal value to you individually, but when aggregated with millions of other users' data, it might become highly valuable for urban planning or targeted advertising. Similarly, in a world where even more substantive forms of data, like songs or feature films, are available in ever-increasing abundance, what consumers are ultimately willing to pay for is not so much the data itself but rather the mechanisms for making it discoverable and convenient to access in bulk. The value of data can also shift dramatically over time, as technologies evolve and new uses are discovered.

While the emphasis on data privacy is vital for protecting individual rights, it also reinforces the notion that data of all kinds is first and foremost private property—and inherently valuable as such. As many private commons scenarios illustrate, however, conceptualizing data as a kind of quasi-public good that is proactively shared, aggregated, and applied in novel ways can help create services where users derive far more value than the platform operators themselves capture.

Universal Networked Intelligence

In the digital world, the tragedy of the commons, if there has to be one at all, does not lie in providing open access to common-pool resources that would then be exhausted through unfettered usage. Instead, it occurs when people try to put limits on how much data we create, how much we share, and who can share it. Unlike pastureland or trees, data is nonrivalrous and ever-multiplying. Capitalizing on those characteristics maximizes value.

This is even more true given the impact that AI will have on the overall value of the private commons. Because of AI's capacity to analyze, assess, retrieve, summarize, and synthesize relevant data from the mountains of data we now generate, the value of the private commons is already increasing. But it could continue to increase by whole orders of magnitude if we embrace an open and collaborative approach to data sharing, while simultaneously developing robust frameworks for privacy protection and ethical data use.

Unlike physical commons, which even with careful stewardship run the risk of depleting over time, digital commons tend to get better. Much better. Both Wikipedia and YouTube are already better resources than when Brynjolfsson and Collis queried users about how much they valued them in 2017. Wikipedia has more articles than it did then. Its articles tend to get more comprehensive and accurate over time as more users contribute to them. For many other private commons platforms, the same dynamics are in effect.

Now think about what happens when you add AI to the mix. Legacy search engines are not very adept at reading our minds. They're largely indifferent to how well they serve us or not. If you enter "jaguar" as a search word, you might get returns for a car, an animal, or a football team. Ask it if any Jaguars own Jaguars, and it will have no idea what you're getting at and no interest in trying to understand you.

With an LLM, it's different. It may not be able to definitively answer your query, but it will certainly have a better shot at it. And we're already starting to see how it will soon be possible to have AI assistants that can successfully act on instructions like "Find me that thing about that Bitcoin podcast I was listening to like maybe six months ago? The one about Operation Checkpoint 2.0? Then summarize the transcript of it, about a thousand words. And then give me links to five solid articles on the topic, from credible sources. Nothing too fringey."

In short, AI will soon act as an intelligent interface layer between you and most, or maybe even all, of the services you use. So, it will increase the value you derive from the many private commons platforms and services that you already use and value. But there's an even more transformational factor to consider. LLMs have now started to go multimodal. That dramatically changes how you interact with them, and how *often* you can interact with them.

When OpenAI demoed GPT-4o, its first true multimodal model, in May 2024, some people responded with disappointment because the model didn't significantly improve its capacity to simulate commonsense reasoning, or solve brain-teasers designed to fool it, or eliminate hallucinations. But what this perspective failed to account for is how dramatically GPT-4o's native multimodal capabilities do change things. (The first versions of GPT-4o released to the public did not yet contain all of the features that OpenAI had demoed. And as we write this, the full range of GPT-4o's multimodal functionality is not yet widely available.)

A fully multimodal model means that you can input any combination of text, audio, images, and video into such models. In turn, fully multimodal models can output any combination of text, audio, and images in their replies. In the past when you spoke to ChatGPT on your phone, it had to translate your audio input to text, process that text, process its response to that text, then convert that text into an audio format to generate a reply to you. This made conversing with ChatGPT different from conversing with a human—not painfully so, but not the same.

With the fully multimodal version of ChatGPT-4o, the translation steps between modes are no longer necessary. Speak something to ChatGPT-4o and it can reply in as little as 232 milliseconds. In addition, it has a much greater capacity to change its cadence, vary its emotional intonation, and adjust other aspects of its vocalizations to make itself considerably more expressive. So conversations with an LLM are no longer just human-sounding. They're starting to become human-paced—with an immediacy and improvisational nature that can transform slightly awkward interactions into breezy BFF banter.

As the ChatGPT-4o launch demos showed, ChatGPT could also now see the world through a user's video camera. Instead of merely uploading a photo for it to analyze, you can point your camera at something in real time and it can instantly process what is happening. In theory, at least, you could take ChatGPT-4o to a movie theater in a foreign country and have it translate a movie in real time. (Please use earbuds if you do that!) You could take it with you to your local golf course and have it analyze your swing. You could use it for fashion advice on combining specific items from your wardrobe, car maintenance advice, plant care guidance, etc.

With these new sensory capabilities, these features have the effect of transforming AI models from a disembodied entity in the cloud that you type to, to an entity in the room that you share space and experiences with. Even if you have an iPhone, your phone is now an android of sorts (the kind with a lowercase "a")—present, in the moment, fully engaged. And more to the point, the app is on your phone, and your phone, rather than your desktop or laptop, is what you truly want to use as your portal to the private commons.

This shift to multimodal models that work on your phone will generate additional consumer surplus. After all, it's our smartphones where we already experience the benefits of the private commons most, because they're always with us, constantly connected, and intimately integrated into our daily routines. Phones are the first thing many people reach for in the morning and the last thing they interact with at night. The smartphone's combination of mobility, connectivity, and multifunctionality makes it an extension of ourselves, always ready to capture moments, answer questions, or connect us to our social and professional networks.

Unlike desktops or laptops, smartphones go everywhere with us, making them ideal for spontaneous interactions with AI, capturing real-world data, and providing immediate assistance in any situation. One of the remarkable things about the reception LLMs have gotten since ChatGPT's release is that they were adopted so enthusiastically even though they mostly functioned as old-fashioned websites in the first two years of their mainstream rollout. Some mobile apps existed but, with audio lag and other factors, LLMs were really only easy to use with a physical keyboard and a monitor larger than the average smartphone screen.

So, how much value gets created when your smartphone starts living up to its name and gets wildly smarter, and also more responsive and capable of interpreting the world with its own sensory awareness? That, combined with an LLM's growing capabilities to act autonomously, will dramatically increase the benefit that users derive from the devices that most of us already love to carry with us. To offer one small example, future driving apps won't just prompt you when to make the next turn. If they sense you're getting sleepy, they might encourage you to take a break or change your Spotify playlist to more high-energy songs.

Or imagine you want to find the best YouTube video, out of the dozens or possibly even hundreds available, on how to play the guitar solo from "Stairway to Heaven." You could have your LLM evaluate the transcripts and comments of all available videos on the topic and recommend the ones it deems most likely to provide the most accurate tabs and best camera angles for learning. More broadly, you could ask it to evaluate thousands of YouTube videos on mastering the guitar and then craft an "Introduction to Guitar Playing" syllabus that gathers the most effective ones.

Advanced AI agents will also be able to draw upon multiple private-commons resources in seamless ways to provide personalized and value-enhancing experiences for individual users. Think about the ways an AI travel assistant could help you manage your upcoming business trip. Once it learns about your trip from its periodic scan of your calendar, it starts making plans. First, it uses maps and Tripadvisor to find you a nice hotel or Airbnb near your meetings and events. Then it looks at your Square history to see which kinds of restaurants you like, maybe even gauging your satisfaction by noticing where you've left the biggest tips. Combining that information with Yelp reviews of restaurants in the destination city, it makes your meal reservations for you. Finally, knowing that you're an avid runner, it checks Strava for popular local routes and cross-references them with your own Strava history to recommend options that are a good fit for your training goals.

The overall effect is not just convenience, but a systemic shift in how people invest their attention and efforts, and what kinds of support and expertise are available to them as they navigate the twenty-first century. We'll explore this topic more in Chapter 6, but for now we simply want to reiterate that AI developers are not extracting or depleting scarce resources when they use massive quantities of data to train new models. Instead, they're engaging in a kind of digital agriculture, or even digital alchemy, that can enable new levels of individual agency and broad-based societal abundance, if we all manage this process effectively.

A popular conceit in the AI development community is that AI is the worst it will ever be right now. As true as that is, we propose a corollary. Even if today's LLMs get no better than they already are, the average twenty-year-old today will arguably reap millions of dollars in lifetime consumer surplus from the LLMs he uses over the course of his life. That's how powerful the private commons becomes in the age of AI.

CHAPTER 5

TESTING, TESTING 1, 2, ∞

In the months after ChatGPT's release, as news media outlets sought to contextualize the scope, speed, and impact of this new wave of AI development, a recurring theme quickly emerged:

> "Prepare for Increasing Model Reliability: The AI Space Race Is On"
>
> "In the AI Space Race for Benchmark Supremacy, Only the Most Accurate and Nontoxic AIs Will Survive"
>
> "Google and Microsoft Now in a High-Stakes Space Race of Documented Performance Increases"
>
> "The AI Space Race Heats Up: How Data Scientists Relentlessly Assess AI's Every Move"

Just kidding. None of those headlines are real. But these are:

"The AI Arms Race Is On. Start Worrying" (*Time*[1])

"Google and Microsoft's AI Arms Race Could Have 'Unintended Consequences,' an AI Ethicist Warns" (CNN[2])

"In AI Arms Race, Ethics May Be the First Casualty" (Axios[3])

"Efforts to Avoid a 'Suicidal' AI Arms Race Are Failing, Scientists Warn" (*Newsweek*[4])

"The Only Winner of an AI Arms Race Will Be AI" (Bloomberg[5])

Very quickly, these early strikes escalated into full-blown mutually assured sensationalism, as hundreds of media outlets engaged in a kind of journalistic arms race to see who could use the phrase "arms race" most often.

But is it really appropriate to equate ChatGPT's capacity to, say, render the Declaration of Independence as a series of mean tweets from Benjamin Franklin to the British Parliament, with high-yield nuclear warheads that can instantly annihilate a city of millions? Does it make logical sense to equate a process that is playing out publicly, with broad global participation, to one that is decided at the highest levels of government, often in extremely secretive fashion?

Certainly there's a bona fide temporal component to AI development. Two-plus years into the era of democratized, hands-on AI, a very real atmosphere of competition defines the terrain. Hundreds of models are available for public use. Hundreds of millions of people around the world regularly use them. This has helped accelerate cycles of deployment, feedback, and steady improvement that map well to the mid-century space race, where, in just a little over a decade, we saw a rapid progression from Sputnik's first orbit of Earth in 1957 to Apollo 11's lunar landing in 1969.

With machine learning and LLMs, we've seen a similarly rapid progression, from DeepMind's DQN mastering old Atari console games in 2013, to AlphaGo defeating world Go champion Lee Sedol in 2016, to AlphaFold making major strides in protein structure prediction in 2020, to today's frontier models that are capable of translating a natural-language prompt in Icelandic into a working computer program, without ever having been explicitly trained on Icelandic or computer programming.

But "arms race," and its intentional connotations of danger and recklessness; innovation as a mad dash to Armageddon? That conceit erases the main attribute that characterizes AI development: eye-glazingly comprehensive testing.

As a rule, AI development does not attract the kind of personality types who go with their gut, play it by ear, or trust their inner voice. Instead, it's a domain largely populated by extreme data nerds who love testing things even more than a TikTok influencer loves seeing a hot take go viral. For them, the "measure twice, cut once" approach to quality assurance is downright slipshod. In their ongoing commitment to data-driven improvement, they measure continuously and cut often.

In fact, tests of artificial intelligence existed before artificial intelligence did. "Can machines think?" British mathematician and computer scientist Alan Turing asked in "Computing Machinery and Intelligence," a paper he contributed in 1950 to a prestigious journal of philosophy called *Mind*. In this paper, Turing laid the groundwork for evaluating machine intelligence through a game in which a computer tries to convince human judges that they're communicating with another person. Today, we call that process the Turing Test.

The Turing Test set the stage for decades of increasingly sophisticated AI assessment. Today, AI developers test model performance in hundreds of different ways. Then they invent more tests to measure the efficacy of those tests. Then they publish white papers documenting their findings. For long stretches over the last seventy years, much of the foundational research and many of the key breakthroughs in AI have occurred in higher education research labs rather than commercial settings—and that heritage reveals itself in AI's robust and data-driven testing culture. Even as commercial developers have come to play a greater role in AI development, this culture of continuous testing and evaluation persists, in ways that foster improvement across the field.

Keep all that in mind the next time you come across an "arms race" story. In the realm of AI, at least, the "race" in question isn't a mad dash or a land grab. It's more like an Ironman triathlon, only longer. The AI models you see today are built on years of carefully administered tests designed to measure their performance across multiple dimensions.

Competition Is Regulation

Outside of Hollywood sci-fi blockbusters, tests were also the most likely way the general public was introduced to AI and its rapidly increasing capabilities. Think of IBM Deep Blue's defeat of world chess champion Garry Kasparov in 1997. Think of Watson trouncing *Jeopardy!* champs Ken Jennings and Brad Rutter in 2011. On the surface, these were entertaining games pitting humans against machines. More importantly, they were tests signifying breakthrough improvements in algorithms, architectures, and data-handling techniques.

Over the last two decades, improvements in machine-learning algorithms and the advent of Big Data have enabled increasingly powerful and multifaceted models. As the capabilities of these technologies expanded, developers devised more tests, or benchmarks, to assess them. Benchmarks have long played a key role in progress throughout the computer industry. Essentially, some organization develops a standardized test for measuring system performance of one kind or another. The goal is to create replicable procedures that yield well-defined metrics on specific tasks. That way, whoever is running the benchmark can compare their results against a previous baseline they've established. Or they can see how they stack up against others in the industry who have run the same benchmark.

Unlike ad hoc testing and other forms of internal validation, benchmarks are typically created by a third party—often an academic institution or industry consortium intending to create a standard around some important aspect of product performance. Thus, when you run a benchmark, you are essentially agreeing to play by someone else's rules to measure and, in effect, objectively certify a given aspect or attribute of your product, be it hardware or software.

For example, in 2019, a collaborative team of researchers from four institutions, including New York University and Facebook AI Research, created a benchmark called SuperGLUE (GLUE stands for General Language Understanding Evaluation). SuperGLUE tests models on eight tasks that are each designed to probe a different facet of language-understanding.

One involves multisentence reading comprehension, in which a model must answer multiple questions based on a short passage. Another task, called word sense disambiguation, tests if a model can determine if a given word has the same meaning across different contexts. A third, called coreference resolution, requires the model to determine the correct referent of a pronoun within passages that feature multiple nouns. Thus, a test sentence might read: "Mark told Pete many lies about himself, which Pete included in his book. He should have been more truthful." To complete the test successfully, the model must recognize that the underlined pronoun ("he") refers to Mark.[6]

Along with providing access to the SuperGLUE dataset and instructions for how to perform the benchmark, SuperGLUE's creators maintain a public leaderboard at SuperGLUE's website. There you can see how dozens of models, tested by individuals and teams affiliated with companies like Microsoft, Google, IBM, Tencent, and Baidu, to name just a few, scored on each of SuperGLUE's eight sections.

Thus, with a combination of collaboration and competition, benchmarks help promote norms of transparency and accountability. As AI researcher Matthew Stewart has noted, they turn development into a "communal Olympics,"[7] where common standards establish both the capabilities of individual models and the overall improvement of AI development over time. You also don't have to be a model's developer to subject it to a benchmark. You can independently evaluate a publicly available model using established benchmarks to assess its performance or limitations.*

So, while benchmarks aren't legally binding like regulations, they do set a standard that many participants in the AI domain strive to match or even exceed. And they can function as a gatekeeping mechanism. Algorithms that perform poorly on benchmarks often get shelved before being considered for real-world applications.

But while benchmarks can function in ways that approximate more formal kinds of regulation, they also offer additional benefits. A regulation is, by design, a fairly static form of governance. It gets drafted,

* Without access to the source code or the training data, you may not be able to assess certain internal aspects of the model such as how efficiently it runs, its resource utilization, or certain safety features that may be implemented at the code level. Nonetheless, many important characteristics like accuracy, reliability, and even some safety and fairness aspects can be evaluated to a large extent.

then it gets deliberated on and revised. Ultimately, the goal is to define, clearly, and with an appropriate degree of precision, what is permissible and what is not. Then the regulation "goes on the books"—where it can often become hard to undo or even update. The longer a law stays on the books, the more likely it is to fall into the trap of governing the present through the lens of the past.

While regulation can be effective for establishing and maintaining a baseline level of quality, safety, or fairness, it doesn't necessarily incentivize improvement. In contrast, benchmarks serve as dynamic mechanisms for driving progress. Sometimes, tests do have explicit governance functions; only physicians who have passed the U.S. Medical Licensing Examination can legally practice medicine. But, generally speaking, the primary role of a test is not to limit, restrict, or otherwise set the scope for allowable behavior. Instead, it aims to evaluate aptitudes or performance. In this way, tests inevitably inspire improvement. Once you know what your score is, you want to beat it. Once you see someone else achieving a certain level of proficiency, you want to equal or exceed it.

So while testing and regulation both aim to standardize and control, testing elevates the focus from compliance to continuous improvement. It's regulation, gamified.

Measure What Flatters?

While usage varies, thousands of benchmarks now exist, providing developers with a wide range of lenses through which to assess their work. Benchmarks that measure accuracy or performance—how well a model correctly identifies images or predicts the next word in a sentence—are the mainstays of this testing, but also just the start.

There are also benchmarks for fairness that attempt to assess whether AI models make equitable decisions across different demographics. There are benchmarks for reliability and consistency; benchmarks that measure a system's resilience against errors and adversarial attacks; benchmarks that assess how understandable or explainable an AI system's decisions are to humans; benchmarks for safety, privacy, usability, scalability, accessibility, cost-effectiveness.

Some benchmarks assess an AI's commonsense reasoning, measuring how well it can make inferences that humans consider obvious based on everyday knowledge. There are also dialogue and interaction benchmarks, which evaluate an AI's ability to engage in natural, context-aware conversations over multiple exchanges.

Within each of these categories, there are distinct subcategories and metrics. Take safety. RealToxicityPrompts evaluates how often language models generate toxic or harmful content in response to certain prompts. StereoSet tests a model's tendency to exhibit various social biases, including ones related to gender, race, religion, and profession. HellaSwag assesses a model's commonsense reasoning by asking it to complete scenarios with plausible endings. The A12 Reasoning Challenge (ARC) tests causal reasoning and reading comprehension, using a dataset of more than 7,000 grade-school science questions.

A benchmark doesn't prohibit a model from engaging in undesirable behaviors—it's just a test—but it does give developers a consistent way to see how much impact they're making with their fixes and adaptations and new approaches to solving shortcomings of their models. Over time, benchmarks can drive major improvement and serve as a public demonstration of that progress as well. In January 2022, for example, OpenAI published a paper about InstructGPT, a precursor to ChatGPT.[8] Note how it invokes the benchmarks TruthfulQA and RealToxicityPrompts to substantiate its progress:

> Compared to GPT-3, InstructGPT produces fewer imitative falsehoods (according to TruthfulQA) and are less toxic (according to RealToxicityPrompts). We also conduct human evaluations on our API prompt distribution, and find that InstructGPT makes up facts ("hallucinates") less often, and generates more appropriate outputs.
>
> Specifically, InstructGPT produced truthful outputs 41.3 percent of the time according to the TruthfulQA benchmark while GPT-3 had produced truthful outputs only 22.4 percent of the time. For RealToxicityPrompts, InstructGPT produced toxic outputs 19.6 percent of the time, a small improvement over GPT-3, which had produced toxic outputs 23.3 percent of the time.[9]

Those numbers may not induce much confidence, but successive models, tested on the same benchmarks, have done much better. GPT-3.5 produced toxic outputs only 6.48 percent of the time on the RealToxicityPrompts benchmark. GPT-4 produced toxic outputs just 0.73 percent of the time.[10]

Benchmarks have played a similar role in measuring and spurring progress across the AI field. There's a benchmark that has tracked the dramatic progress in models' ability to recognize objects. A popular machine-translation benchmark called BLEU (Bilingual Evaluation Understudy) offers a simple numerical measurement to assess how accurate Google Translate is for different language pairs, like English to French or German to Spanish. The Word Error Rate (WER) benchmark has been instrumental in highlighting declining error rates for voice assistants like Amazon's Alexa and Apple's Siri.

As it turns out, a truly effective benchmark can optimize itself into obsolescence, by inspiring performance gains so great the benchmark no longer poses a sufficient challenge to the models it was designed to measure. According to the 2023 *AI Index Report*, published by Stanford's Institute for Human-Centered AI, "performance saturation across many popular technical performance benchmarks" now characterizes the field.[11]

As the *AI Index Report* notes, researchers have responded to this challenge by trying to create more innovative benchmarks that test models in more comprehensive and complex ways. Yet even as benchmarks have helped drive LLM improvements in ways that enable increasingly impressive performance in real-world scenarios, it's also true that the multiple-choice questions that real life throws at you are never quite as tidy as the ones that appear on exams. So, a fundamental argument within the AI community persists. Are today's most proficient models really making any progress toward more generalizable and humanlike intelligence? Or are they just scoring higher, through benchmarks that don't do enough to genuinely measure true understanding and adaptability? The real goal of a test, after all, isn't just to confirm that test takers know the correct answers, but also to show that they've acquired capabilities and expertise that they will be able to apply in a range of settings.

We're Going to Need to Build a Bigger Test

As in the classroom, so in the lab: in AI, *teaching to the test* is a thing. Especially when models were smaller, and benchmarks tended to be narrower in scope, researchers often trained their models on data that was very similar to a target benchmark's dataset, often through supervised learning.

Developers training a model to do well on an object-recognition benchmark might provide it with thousands of labeled images that showed a wide range of objects from various angles and in different lighting conditions.

Through such training, models did get dramatically better at recognizing objects. In 2015, a computer vision model called ResNet won an annual benchmark competition known as the ImageNet Large Scale Visual Recognition Challenge (ILSVRC) with a score that surpassed even human abilities. This marked the beginning of a new era. Now computer vision models routinely outperform human accuracy in specific visual tasks, like facial recognition and medical image analysis.

But models can still make mistakes that reveal they don't necessarily have any greater conception of the physical properties of a given object, or how objects relate to each other in the real world. Their susceptibility to what's known as "adversarial examples" helps illustrate this. Change a few pixels in an image that a model correctly identified as a zebra, and the model might then insist the image is a toaster.

This, of course, is an inference that even a nearsighted toddler would be unlikely to make, because for the most part we humans understand that a single object is not likely to straddle the line between being a zebra and being a toaster. To us, those two things just aren't all that much alike. Our mental model of the world and store of common sense readily steer us away from that proposition.

As all kinds of models evolve, their vulnerability to adversarial inputs and other kinds of error generally diminishes. Where GPT-2 once struggled to maintain coherence over long passages, for example, GPT-4 can now produce consistent and logically structured content across thousands of words. In the face of such performance increases, researchers and developers devise increasingly complex benchmarks in an effort to discern if today's frontier models are truly acquiring new cognitive capabilities that go beyond memorization or sophisticated pattern-matching.

In April 2023, for example, a team of researchers from Microsoft published a research paper[12] that summarized their efforts to create novel questions and tasks that could challenge GPT-4 to display capacities for interdisciplinary problem-solving, visual reasoning, and other cognitive abilities associated with general intelligence. In one often-cited example, these researchers asked it to draw a unicorn using a programming language called TikZ that is typically used to produce charts, diagrams, and other vector-based graphics through command-line instructions rather than through the kinds of graphic user interfaces found in programs like Illustrator or Photoshop.

What made this task so difficult? Think of what it would take for a human to fulfill this challenge. First, he'd need to know how unicorns are generally depicted. Then, because TikZ uses mathematical terms expressed through programmatic instructions to produce precise geometric shapes and lines, he would have to figure out how to approximate the visual appearance of a unicorn in light of these characteristics. He'd also have to be familiar enough with the syntax and functions of TikZ to use it in a somewhat novel way—drawing a figurative illustration rather than using it to produce a well-established graphic form like a bar chart or Venn diagram.

To do this successfully, he would have to use abstract thinking and problem-solving to break down the image of a unicorn into separate programmable components: a snippet of code to represent the horn; another snippet to represent its mane. He'd need to possess some spatial reasoning to effectively position and scale the various elements of the drawing within the coordinate system used by TikZ, which defines the drawing space as a grid with horizontal (x) and vertical (y) axes, much like a map or graph paper. All in all, it would take both knowledge across multiple domains and the creativity to synthesize that knowledge in appropriate ways to achieve the end result, a two-dimensional composition of geometric shapes and lines that is recognizable, at least conceptually, as a unicorn.

So how did GPT-4 manage this challenge? Over the course of a month, as OpenAI was continuing to refine GPT-4, the Microsoft researchers asked it to draw a unicorn with TikZ three times. On its first try, GPT-4 produced an image so rudimentary only a parent would have hung it on the refrigerator. But this image did incorporate a number of features typically associated with unicorns, if you knew what you were looking for. It was pink, and it had a nominal horn and a bushy tail. And GPT-4's two subsequent efforts were considerably more identifiable. The unicorn's mane and tail grew more pronounced. The shape of its head got more equine. Its overall proportions displayed more balance and cohesion.

As this and other tasks demonstrate, state-of-the-art models now regularly engage in feats that, at the very least, *seem* to go well beyond pattern recognition. They can often explain their decisions and actions in ways that suggest a sophisticated grasp of human intentions and emotions. They can summarize and synthesize information with a competence that approximates considerable degrees of comprehension.

But they also continue to make mistakes that suggest a lack of true understanding that generalizes across different domains. Instead, they are arguably just getting even more adept at increasingly advanced kinds of pattern-matching. "A machine can pick up on very subtle statistical associations in the language that humans would never pick up on," explained Melanie Mitchell, the Davis Professor of Complexity at the Santa Fe Institute, in *Nature*.[13]

By now, some Big Tech developers have begun to train their models on datasets containing a trillion words or more. When datasets get that big, skeptics contend, it's no wonder that an AI can expertly pair any New Zealand sauvignon blanc with a dish that complements the varietal's pronounced fruit flavors, because ample tasting notes for New Zealand sauvignon blanc and other relevant information are part of its data. In effect, this argument goes, these models are "cheating." They've seen the answers before they take the test.

This phenomenon is known as data contamination or data leaking, and developers go to great lengths to prevent it from happening. If a model has been inadvertently exposed to its test data during training, it can lead to artificially inflated performance metrics and an inaccurate assessment of the model's true capabilities. Whatever short-term publicity and reputational incentives may exist for gaming the system and achieving a high score simply for the sake of achieving a high score, most developers are far more focused on making progress toward genuine, generalizable intelligence with real-world utility.

Since benchmarks, used correctly, are a useful proof of progress, developers try to maintain strict separation between the data they use to train their models and the data they test their models on. They implement robust tracking measures to understand the sources of the data they aggregate, and conduct various tests to detect potential contamination. They rigorously exclude known test sets from training data.

Because they take such precautions, improvements on benchmarks do in fact reflect genuine advances in AI performance. At the same time, hallucinations and the other kinds of nonsensical or factually incorrect outputs that models produce continue to serve as strong rebuttals to any claims about human-like intelligence. So, if LLMs are truly getting more capable in more generalizable ways, why do they keep making the same kinds of dumb errors?

This persistent capacity for error, Gloomers assert, makes models brittle. At a crucial moment, an LLM that has aced medical licensing exams and can recite complex diagnostic criteria still might not pick up on the subtle contextual clues in a patient's description of their symptoms—potentially leading to a missed diagnosis of a time-sensitive condition like early-stage sepsis or a mild stroke.

That's true as far as it goes; we certainly haven't reached a point where models literally never make errors. We may never reach that point. But if our goal is progress rather than perfection, do we need to? Humans, after all, make errors, lots of them. It's one of the main reasons we look to machines in general, and certainly AI models, as means for improving human capabilities.

So how certain do we need to be of LLM performance to trust it, and how do we get there? Regulation is one way we try to compel certainty, but no regulation can completely eliminate the risk of some unfortunate thing happening. Laws that make robbery a crime aren't a guarantee that you won't ever get mugged—they're simply a policy designed to reduce that possibility. Attorneys and doctors must demonstrate their expertise before obtaining licenses to practice their disciplines, but that doesn't mean a surgeon won't mistakenly amputate a patient's wrong leg. It's happened.

Now, if that were to happen to you, it probably wouldn't be because the surgeon didn't really have a good understanding of what legs are, that there are right ones and left ones, and that they're not interchangeable, especially in surgical contexts. It would probably be because someone made a mistake on a chart, or that the surgeon was drunk, or something else of that order.

Ultimately, though, what matters more: how or why an error is made, or how often such errors happen?

AI researchers, developers, and ethicists often emphasize the importance of two related concepts: model interpretability and explainability. Interpretability focuses on the degree to which a human can consistently predict a model's results. The more inherently transparent a model's structures and inputs are, the easier it is for humans to accurately anticipate its outputs. Explainability refers to the *how* of a model's decision-making processes: Is it possible to explain, in broadly understandable terms, how a system decides that an image contains a cat or that a particular financial transaction is fraudulent? In essence, explainability aims to demystify the "black box" nature of AI decision-making, often after the fact.

These concepts are seen as crucial for building trust in AI systems, particularly in high-stakes domains. And yet while developers should certainly strive for interpretability and explainability, establishing *absolute* interpretability and explainability as the standard for "safe" AI would be unrealistic, unproductive, and highly unusual in the larger scope of how the world works.

In our legal system, in both civil and criminal contexts, we sometimes take circumstances and motives into account, but actions are always at the heart of a matter. With AI, the same should be true. At least if your primary intent is practical AI adoption rather than AI prohibition, *how* a model does something is important, but not as important as *what* it does.

A model's capacity to make decisions and generate outputs at scale is a crucial aspect of what it does, and therefore *should* be a factor in how much we trust it. But Gloomers often exploit this fact by presenting various instances of edge-case anecdata as evidence of inherent systemic failure: If a computer vision model can mistake a toaster for a zebra because the model has no real understanding of the world—and you know this will happen, at least occasionally, as widespread adoption occurs—then how can you trust it ever? If it's repeatedly stumped by certain kinds of logic problems, doesn't that make it fundamentally unreliable for all tasks that require a capacity for genuine reasoning or understanding?

Zoom out, though, and what gets presented as an inherent limitation can begin to look like a statistical anomaly. Instead of demanding perfect performance—an unrealistic standard we don't apply to humans—we should focus on establishing acceptable error rates and continuously improving overall system reliability. Just as we trust human-driven systems despite known error rates, we can develop trust in AI systems that demonstrate consistent, measurable reliability.

Writing in *Science*, Jack Stilgoe, a professor of science and technology policy at University College London, similarly argues for evaluating AI from a public perspective, emphasizing societal utility over technical capabilities:

> For society to make good decisions about AI, we should instead look to another great late-twentieth-century computer scientist, Joseph Weizenbaum. In a paper "On the Impact of the Computer on Society," in *Science* in 1972, Weizenbaum argued that his fellow computer scientists should try to view their activities from the standpoint of a member of the public. Whereas computer scientists wonder how to get their technology to work and use "electronic wizardry" to make it safe, Weizenbaum argued that ordinary people would ask "is it good?" and "do we need these things?" As excitement builds about the possibilities of generative AI, rather than asking whether these machines are intelligent, we should instead ask whether they are useful.[14]

While Stilgoe seems fairly skeptical regarding generative AI's utility, the questions he's asking, by way of Weizenbaum, are ones we certainly should be asking. To determine if something is good, we have to know something about its nature. To know if it could be useful to us, we have to know something about its utility.

Together, both of these essential questions prompt more questions. What are the intentions of the thing in question—in this case, generative AI? What are its stated values? How does it uphold them? What is its impact on the world, ethically, culturally, environmentally? Do its impacts align with your own values? Do you trust it?

Benchmarks help researchers and developers explore these questions and gather a deeper understanding of an AI system's capacities, but what resources are available to the "ordinary people" whom Weizenbaum identifies as the key evaluators of a new technology? As it turns out, there's a way to combine some of the testing dynamics that inform the practice of benchmarking in a format that's wide open to the general public.

Are You Not Infotained?

Chatbot Arena is an open-source platform for evaluating LLMs based on human preferences.* It works just like ChatGPT or Claude: type your prompt into a text box, submit it, and wait for the reply. The difference here is that your prompt is simultaneously submitted to two LLMs, neither of whose identity you know. You compare the output each one generates, and vote for the one you like best.

Based on all the votes cast in these head-to-head showdowns, Chatbot Arena's administrators stack-rank all the models on the platform's leaderboard. In a process similar to how college football or PGA rankings are determined, the administrators use sophisticated mathematical methods to establish which models are the best overall, even if specific

* You can access Chatbot Arena at https://lmarena.ai/. The website was developed by members from LMSYS, an open-research organization founded by students and faculty from UC Berkeley and UC Berkeley SkyLab.

models never compete against each other. As of this writing, the site's leaderboard ranks more than 130 models.

The limited scope and controlled conditions of traditional benchmarks, optimized for apples-to-apples comparisons, means they can't provide a truly comprehensive picture of how a model will perform in broad, open-ended, messy, and often rapidly evolving real-world conditions.

In contrast, Chatbot Arena drives improvement through a single all-encompassing metric: general customer satisfaction. In this respect, the leaderboard resembles many of the internet's most effective governance mechanisms, like eBay's star ratings or Reddit's upvote system, which distill complex interactions into simple, universally understood signals.

In "Regulation, the Internet Way," a 2015 essay published by the Harvard Kennedy School's Ash Center for Democratic Governance and Innovation, Nick Grossman, a general partner at Union Square Ventures, observed that "the central inversion in internet-style regulation is the flip from 'permission' to 'accountability.'"[15] What he meant was that internet platforms and marketplaces, like eBay and Airbnb and Uber, didn't require would-be participants to obtain business licenses, or seller's permits, or any other kind of bureaucratic certification as the up-front price of participation.

Instead, you could jump right in. But once you did jump, you didn't actually land in some new realm devoid of regulation. In fact, the internet was and is a highly regulated space, where billions of routine transactions and interactions get scored, aggregated, analyzed, and transformed into reputation scores and other transparency and accountability metrics that perform governance functions that are flexible enough to keep up with the speed and the scale at which the internet moves.

And that's what the Chatbot Arena leaderboard does for AI models. Unlike many of these kinds of internet governance mechanisms, which evaluate participant behavior, Chatbot Arena evaluates system outputs. Aggregated over time, individual votes create collective "rulings" on both individual model performance and, more generally, how the public believes models should be functioning.

As we'll discuss in more detail in the next chapter, one key aspect of OpenAI's iterative deployment approach is how it enables decentralized hands-on testing at a scale you could never achieve in a lab. Hundreds of millions of people around the world are now using hundreds of models. As this process occurs, each of these users is learning individually about how models operate, what they personally value and trust, and what they don't value and trust. Developers are learning these things too, at scale, for their own models. They can see what people like and don't like, and where issues are arising, and can then incorporate their findings into future updates and releases.

But leaderboards like Chatbot Arena are designed to aggregate data at a larger scale, tracking performance data for 100-plus models at once, not just one. With that, users can learn from the experience of other users, not just from their own experience. And this collective intelligence can be relatively easily applied to all participants in a sector, which is something new.

For example, there's no easy way to determine average user satisfaction for how YouTube or Facebook are performing overall, much less compare them in ways that would provide a clear indication of their quality. The same is true for news media publications: there's no site that lets you rank, blindly, how three different outlets handled the same story. And even if there were, such a site would be easy to game. You could search to find which publication ran which story before casting your vote.

But at Chatbot Arena, the models generate their responses in real time; their outputs only exist because you just prompted them into existence. So it's generally resistant to attempts at gaming the system. As Chatbot Arena's administrators continue to develop the site, it's also offering more granular forms of assessment. As a default, it shows you an overall ranking that indicates which models have received the most winning votes from users. But the site can evaluate data more finely than that. For example, it can group the kinds of prompts users are making into various categories, such as "Instruction following," "Coding," "Math," or "Longer query," then show how a model's performance may differ for different kinds of use cases. A model that is ranked highly for math, for example, may not perform as well when responding to longer queries.

This is another instance where more usage will lead to more functionality. Imagine, for example, if a few hundred million people started using Chatbot Arena to evaluate model performance. With that many people submitting prompts every day, even fairly esoteric use cases might quickly begin to achieve statistical relevance. In theory, users could eventually get one-stop insights into virtually any facet of model performance. Which models are best at creative-writing tasks? Which ones are best at translating from English into Mandarin? Which ones score highly for both "low hallucination rates" and "outputs requiring emotional intelligence and empathy"?

In addition, Chatbot Arena's administrators could run their own tests on the outputs, and get real-world, industry-wide rates for phenomena like factual inaccuracies or toxic outputs. They could ask users why they preferred one output over another, to learn more about user preferences and decision-making processes in AI interactions.

With that kind of potential, Chatbot Arena points the way toward a future approaching the democratized, grassroots governance of Regulation 2.0, in which users end up shaping ongoing AI development through collective expression of their preferences and judgments, and trust is achieved through transparency. If users can easily determine that, say, model X generates false information three percent of the time when giving medical advice, they can make much better-informed decisions about which AI models to trust for different tasks. Developers, in turn, would have major incentives to improve their models in ways that genuinely matter to users.

CHAPTER 6

INNOVATION IS SAFETY

As AI models grow more powerful and capable of doing more things, effective safety measures and broad human oversight grow more critical too. Consequently, the need to make sure innovation is balanced with prudence also grows more critical—but how do we do that?

Gloomers tend to associate safety with attributes like caution, deliberation, and attentiveness. But the pace of development matters too. Regulations and other policy interventions designed to delay development don't just hamper our capacity to create breakthrough products and services that increase global well-being. They can also delay the positive effects that innovation can have on safety.

It's also important to always consider the global context of technology. While American companies and academic institutions remain at the forefront of the machine-learning renaissance that started in the early 2010s, China, France, Israel, India, South Korea, and many others are all determined to use AI to advance their strategic priorities. In a highly networked world characterized by increasing levels of interdependence, maintaining what former Google CEO and Alphabet executive chairman Eric Schmidt describes as America's *innovation power* emerges as a key safety priority. By maintaining our development lead, we're infusing AI technologies with democratic values, and integrating these technologies across society in ways that bolster our economic power, our national security, and our ability to broadly project our global influence.

Rapid development also means adaptive development. And adaptive development means shorter product cycles, more frequent updates, and safer products. Especially in the domain of internet software, where distribution is instantaneous and feedback loops are tight, you can typically innovate to safety much more effectively than you can regulate to safety.

But that's not how Gloomers and Doomers see things. In the first few months after ChatGPT's release, a new sense of urgency informed calls to regulate development and limit public access to these new resources. For example, U.S. Congress member Ted Lieu, a Democrat from Southern California, wrote in a *New York Times* op-ed that he was "freaked out" by "unchecked and unregulated" AI. In his estimation, we need to create a new federal agency exclusively dedicated to this technology, an FDA for AI, to draft and enforce regulations faster than Congress would be likely to manage.[1]

Lucy Powell, digital spokesperson for the UK's Labour Party, suggested that AI development should be regulated like medicines and nuclear power, with licenses required to build models like ChatGPT.[2] Alexandra van Huffelen, the Netherlands' minister for digitalization, drew a connection between AI and automobiles. "Whenever you produce a new car, you first want to make sure that it's safe to drive before you allow it to be in the streets," she remarked.[3]

All of these lawmakers and government officials were arguing for a regulatory approach characterized by what is often described as the *precautionary principle.* In essence, the precautionary principle holds that new technologies are "guilty until proven innocent."[4] Conversely, many technologists, entrepreneurs, and investors favor an approach known as *permissionless innovation.* In contrast to top-down efforts to preemptively control or even prohibit new technologies, a permissionless innovation approach explicitly establishes ample breathing space for innovation, experimentation, and adaptation, especially when tangible harms don't exist yet, or when existing regulations already cover those that do.

As useful as the concept of permissionless innovation is for creating the conditions that help facilitate trial-and-error experimentation, it has an unfortunate consequence. By conceptualizing innovation as something that occurs most productively when it is not unduly constrained by formal regulatory efforts—which is certainly true—we can end up discounting the inherent regulatory power of innovation itself.

We tend to think of regulators and entrepreneurs being diametrically opposed to each other. In reality, both share the common desire to improve how some aspect of the world works. Regulators try to do it by creating new laws. Entrepreneurs do it by creating new products and services that provide novel capabilities or somehow refine existing designs.

In the early days of the automobile, for example, manufacturers had natural incentives to make cars more reliable and easier to use, which in turn made them both safer and more marketable. In this way, every new iteration of a component or feature could be seen as a "law" or "regulation" that the public could then ratify through its adoption.

Take electric starters, which Charles Kettering invented in 1911 and Cadillac added to its models in 1912. Prior to their availability, drivers had to use a hand crank to manually rotate the engine's crankshaft as part of the sequence of steps required to start the vehicle. Not only did the process take a degree of strength that only some drivers possessed, but it also resulted in a fair number of broken wrists, arms, and even jaws, when engines backfired and sent the cranks spinning backward.

The introduction of electric starters, which simply required a driver to press a foot pedal to start the engine turning, helped democratize access to automobiles and made them safer and more trustworthy. Drivers loved them, and as a consequence, electric starters quickly became standard across the industry. By 1920, hand-cranked engines had all but disappeared.

Now, because of how the internet broadens user bases and invites participation, the regulatory function of permissionless innovation is far more powerful than it has ever been. In the twenty-first century, anyone advocating for the freedom to experiment openly and iteratively, without prior approval from official regulators, is not escaping scrutiny and democratic oversight so much as courting it. Ultimately this leads to more comprehensive and inclusive feedback, shorter product cycles, and faster improvements, which generally leads to safer and more trustworthy products.

To understand why this dynamic is so important in the context of AI, let's look at both the precautionary principle and permissionless innovation more closely.

Move Fast and Brake Things?

"New technologies can bring mankind great benefits, but they can also cause accidental harm," Michael Pollan, the bestselling writer known for his work on the food industry and sustainable agriculture, wrote in a 2001 essay for the *New York Times*. "How careful should society be about introducing innovations that have the potential to affect human health and the environment?"[5]

Pollan was writing about the precautionary principle when it was starting to catch on in the U.S. as an alternative to more quantitative forms of risk analysis. As Pollan observed, quantitative risk analysis is effective in some scenarios but less so in others. Because it depends on past data to make future predictions, it isn't always effective at capturing the potential risks of emerging technologies where substantive data regarding impacts does not yet exist. "Whatever can't be quantified falls out of the risk analyst's equations, and so in the absence of proven, measurable harms, technologies are simply allowed to go forward," Pollan wrote.

To address this dynamic, lawmakers in the former West Germany pioneered a different approach in the 1970s. In response to the acid rain that had started to degrade the country's forests, the government passed laws to reduce sulfur dioxide and nitrogen oxide emissions from factories and power plants even though there was not yet definitive scientific proof that such emissions were causing the acid rain.[6]

In West Germany, this approach came to be known as *Vorsorgeprinzip,* which translates to the "forecaring principle," or, if you prefer more cease-and-desist bite, the precautionary principle. Either way, it shifts the burden of proof to the party introducing potential risks. This, Pollan noted in his article, established it as "a radical challenge to business as usual in a modern, capitalist, technological civilization." Instead of prioritizing upstarts, unconventional approaches, and the freedom to innovate, the precautionary principle is fundamentally conservative in legislative temperament. As Pollan phrased it, it's a "better safe than sorry; look before you leap" approach.

Reflecting its environmental roots, the precautionary principle's main impacts in the U.S. have been felt in environmental and public health contexts. Regulations imposed on genetically modified organisms (GMOs) in agriculture often invoke the precautionary principle, requiring extensive testing and labeling even in the ongoing absence of conclusive evidence of harm. Similarly, the restrictions on the use of certain chemicals, such as BPA and neonicotinoids, were motivated by concerns about their potential long-term health and environmental effects. (To date, definitive scientific proof that these chemicals cause harm has not yet been produced.) But the impact of the precautionary principle extends to the technology industry as well. The EU's implementation of the General Data Protection Regulation (GDPR) in 2018 takes a precautionary approach that places a number of default limitations on how organizations can make use of data generated by the individuals who use their services, without prior proof that those uses cause harm. Because organizations from around the world serve customers in the EU or process EU residents' data, the GDPR has a global impact on data practices, including in the U.S.

Precautionary thinking has likewise had an impact on many facets of AI development, especially at local levels. In 2017, for example, a San Francisco city supervisor, Norman Yee, initiated a ban against delivery robots. "For me to wait for something to happen is silly," he told *Wired*. "Because I think it's going to happen."[7] A few months later, the city passed legislation that prohibited food delivery robots in most parts of the city. In the few places they were allowed, they could travel no faster than 3 mph and had to have a human chaperone within thirty feet of them at all times.[8]

In 2020, Portland, Oregon's city council banned law enforcement agencies, the local airport, and private businesses from using facial recognition technology. In 2021, New York City's Department of Transportation issued a rule that required autonomous vehicle companies to self-certify that their cars could "operate more safely than human drivers in New York City" before obtaining a permit.[9] That same year, Virginia prohibited all of its government agencies, including law enforcement, from using facial recognition technology.

Still, the perception persists that the internet is a Wild West of dangerously laissez-faire development, with no sheriff in sight. While this viewpoint underacknowledges the wide range of precautionary interventions that have decelerated tech progress in various ways, it also erases government regulators from the larger story of tech's positive social impacts. In this view of the world, the tech industry's emergence as a global engine of cultural transformation and economic growth kind of just happened, through a fortuitous combination of steady and spectacular increases in computer processing power, free-market magic, a handful of coding superstars, and regulatory indifference.

But that's not true. As public policy analyst Adam Thierer explains in his 2016 book, *Permissionless Innovation,* the conditions that helped make high tech America's most dynamic industry over the last thirty years were actually quite intentional:

> Specifically, beginning in the early 1990s, a bipartisan group of policymakers gave innovators the green light to let their minds run wild and experiment with an endless array of exciting new devices and services. US policymakers signaled that permissionless innovation would be the norm for the Internet and digital technology in America through a series of policy statements.[10]

In 1991, the National Science Foundation, the federal agency that managed NSFNET, the network that formed the backbone of the internet in the U.S., started relaxing its restrictions against its commercial use. This policy shift, occurring over several years, enabled growing private-sector involvement and investment in internet infrastructure and services, which in turn accelerated the growth and accessibility of the internet for personal use. Then Congress passed the Telecommunications Act of 1996, including a provision known as Section 230. Often described as the "twenty-six words that made the internet," Section 230 protects website operators and other online intermediaries from liability claims arising from the content that third parties post on their sites, platforms, and networks.

Without Section 230, popular bloggers like the historian Heather Cox Richardson or the economist Tyler Cowen would theoretically be much less likely to allow user comments on their websites. Developers like Substack, Medium, and WordPress would be much less likely to build blogging platforms for individual bloggers. Moving further up the chain, cloud service providers like Amazon Web Services or Microsoft would be much less likely to host blogging platforms. Consequently, the internet would be far more broadcast-oriented than interactive, with legal and technological barriers to entry that would likely relegate 99.9 percent of the world's population to participation only as shoppers and passive audiences for news and entertainment.

Finally, in 1997, President Bill Clinton and Vice President Al Gore released the "Framework for Global Economic Commerce," a policy document that the *New York Times* described as a "hands-off, no-new-taxes approach to regulating business transactions on the worldwide computer network."[11] Recognizing the internet's emergence as "an appliance of everyday life," the Framework advocated for market-oriented approaches while repeatedly discouraging new and unnecessary regulation, new bureaucratic procedures, and other top-down interventions. "Where governmental involvement is needed, its aim should be to support and enforce a predictable, minimalist, consistent and simple legal environment for commerce," it noted.[12]

Together, these policy choices created an environment for innovation, investment, and remarkable generativity that rippled through the American economy over the course of the 1990s. From 1993 through 1997, the tech industry created more than one million new jobs.[13] By the late 1990s, when the dot-com boom was well underway, GDP was growing at an average annual rate exceeding 4 percent. The unemployment rate dropped from just under 8 percent in 1993 to less than 4 percent by 2000.

In the decades that followed, the permissionless innovation approaches implemented in the 1990s continued to deliver new breakthrough technologies, including cloud services, smartphones, and social media. However, as these technologies became ubiquitous, public sentiment began to shift. In part because of how we tend to habituate to a technology's benefits and focus more attention on its downsides over time, users and regulators alike began to scrutinize the power and influence of tech giants more closely.

In March 2018, grassroots movements like #DeleteFacebook gained traction in the wake of the Cambridge Analytica scandal, when a whistleblower revealed how Facebook had allowed an app developer to harvest personal data from up to 87 million people and use it for political purposes without their consent. Less than a month later, senators summoned Mark Zuckerberg to Capitol Hill, where they picked up right where the lawmakers from the 1960s hearings on the national data center had left off, positioning Facebook and other social platforms as an existential threat to personal privacy that would presumably compromise or even eliminate their ability to shape their own destinies.

Today, concerns over data privacy, disinformation, and other tech-driven consequences have grown so strong that calls for precautionary approaches are becoming more commonplace. But permissionless innovation is arguably working more effectively than ever. Not only did it enable the progress we've seen to date in AI, but it also helped accelerate advances in electric vehicles, CRISPR gene-editing, grid-scale energy storage and solar technologies, cryptocurrency, telemedicine, 3-D printing, and augmented reality.

Innovations like these aren't just improving our daily lives. They keep the U.S. at the forefront of solving global challenges, from climate change to health care accessibility. At a time when moving quickly in a global landscape has rarely been so consequential, and we have a distinct competitive advantage thanks to our innovation power, regulators and anti-tech critics are insisting, in the strongest possible terms, that it's time to move fast and brake things.

The Future of Life Institute's letter calling for a total pause on AI development took this perspective to its illogical extreme. "These [shared safety] protocols should ensure that systems adhering to them *are safe beyond a reasonable doubt*," it proclaimed (emphasis ours).[14] In doing so, it neatly flipped the criminal law standard that guilt must be proven "beyond a reasonable doubt."

In criminal law, the penalties imposed often involve physical incarceration: a jury can literally take away a person's freedom and autonomy. So, our obligation to the accused, to those who were victimized by the crime in question, and to the integrity of the entire criminal justice system is to be as certain as reason allows that the accused actually did what prosecutors are charging them with. In the Future of Life Institute's bizarro-world version of "beyond a reasonable doubt," they advocate locking up a technology because we can't be sure it won't do something bad, someday, maybe.

This is not just predictive policing without empirical evidence. It's predictive sentencing without empirical evidence. In any era, that's an odd fit for a democratic society. In the internet age, it feels even more misaligned. As Michael Pollan noted when describing the precautionary principle's origins in the 1970s, it was a response conceived due to a lack of data. Now, however, our capacity to generate data instantly, share it widely, and analyze it continuously is dramatically more powerful. And, of course, this is even more true when dealing with digital technologies.

This doesn't mean there are no longer scenarios where the precautionary principle makes sense, even with digital technologies. But there's also a reason why concepts like "minimum viable product," "product/market fit," and "going agile" have defined internet software development for the last thirty years.

Learning happens much quicker now. Updates are much easier to push out the door. Freed from the costs, constraints, and slow pace of physical production and distribution, you launch a prototype and discover through clicks, time stamps, comments, and other data what your users find useful, what they dislike, and what they don't care about. Then you use this knowledge to revise and deploy the next version of your product in a very efficient feedback loop of continuous improvement.

Iterative deployment supplements this paradigm with the explicitly prosocial component of measured distribution. The average software developer moves at the pace of their own operational limitations—how quickly can they make the proper inferences from the data they collect? How quickly can they write and test the new code that actualizes their learnings? In deference to the unique cultural implications of synthetic intelligence, OpenAI factors collective adaptation into its distribution decisions. Here's how it described its approach in a 2023 blog post:

> Crucially, we believe that society must have time to update and adjust to increasingly capable AI, and that everyone who is af-

> fected by this technology should have a significant say in how AI develops further. Iterative deployment has helped us bring various stakeholders into the conversation about the adoption of AI technology more effectively than if they hadn't had first-hand experience with these tools.[15]

This approach now extends far beyond OpenAI, as other AI developers quickly followed its lead. The widespread hands-on usage that iterative deployment enables hasn't been risk-free or without negative consequences. But from a broad safety perspective, abandoning this approach wouldn't just inhibit usage, it would inhibit learning and progress. Every day, iterative deployment means millions of people from around the world are testing AI in real-life scenarios. We're learning more, faster, about the most common kinds of usages and issues. We're learning more about what happens when AI systems interact with people with varying levels of technology literacy, from diverse cultural contexts, and in unexpected real-world situations.

We're also still in the early stages of global AI development and adoption. In a few years, when twice as many people are using AI far more often than they are now, and the AI itself is considerably more powerful, will permissionless innovation and iterative deployment still make sense? Will we need to resort to more precautionary approaches?

The past may hold some clues. When Gloomers insist that we should regulate AI the same way we regulate automobiles, they're more right than they may realize. To understand the ramifications of this more clearly, let's take a look at how permissionless innovation was functioning in the early 1900s, in an era that helped set the course for American prosperity and individual agency for the next hundred-plus years.

Road Test

For billions of people over the course of the twentieth century, cars were undoubtedly the most powerful machines they had ever laid hands on. And not just the most powerful machines, but also ones they used multiple times a day, in life-changing ways and trivial ones, for decades. As the historian John B. Rae noted in his 1971 book, *The Road and the Car in American Life*, the rapid adoption of cars in the early decades of the twentieth century afforded "personal mobility of a completely novel kind and on a completely novel scale."[16] As Rae emphasized, the transition wasn't limited to a narrow class of privileged individuals enjoying access to a new technology that amplified the range and speed of personal movement in genuinely extraordinary ways. On the contrary, within just a few decades, millions of people possessed this bona fide superpower. A new era of automobility ensued, completely reshaping long-standing conceptions of individual freedom and patterns of living, and changing the arc of millions of lives and society itself.

Networked PCs and smartphones have been the twenty-first-century version of cars. They're personal mobility machines that provide a new way to project ourselves through time and space. In the AI era, as we endow them with synthetic intelligence, PCs and smartphones become even more powerful vehicles for expressing our agency and exerting our will upon the world. We are in the midst of another epic transition, going from bicycles of the mind, as Steve Jobs once described personal computers, to Ferraris of the mind.

Naturally, new risks will emerge alongside new capabilities. Instead of settling for nothing less than risk-free models, however, we should make it our goal to understand the risks that occur in real-world conditions and systematically work to manage and reduce them. Iterative deployment is how you do that.

A little over a century ago, when the U.S. had a population of 97 million and the Ford Motor Company was arguably the world's fastest-growing high-tech company, there were only around 1 million cars on the road. In 1913, in what was then the world's largest manufacturing plant, a four-story building in Highland Park, Michigan, known as the "Crystal Palace," Ford implemented moving assembly lines for the first time. This new approach reduced production time for a single Model T from about 728 minutes to 93 minutes.[17] Major price drops followed, from $690[18] for 1912's most popular model to $360 for the 1913 version.

By 1916, Ford was producing 577,036 cars annually, an increase of nearly one hundred times from when the company introduced the Model T in 1908. By 1925, the Ford assembly line was completing a new car every ten seconds. As one Ford historian put it, these breakthroughs in productivity "put the nation on wheels."[19] In the span of a few years, Henry Ford's vision of a basic but reliable automobile that everyday people could afford transformed the car from a luxury good for rich hobbyists to a literal engine of progress that radically amplified individual agency and society-wide productivity.

But, of course, the democratization of automobility brought significant harms too. From 1913 through 1923, the rate of motorist-related fatalities per 100,000 people more than tripled. On one day in 1927, eight children were killed in separate automobile collisions in and around New York City. In the first four years after the end of World War I, the book *Fighting Traffic* notes, "more Americans were killed in automobile accidents than had died in battle in France."[20]

What if, in 1913, President William Howard Taft, sensing danger ahead, had stepped in and signed an executive order forbidding automobile manufacturers from producing cars that could go faster than 25 mph?* What if the U.S. Congress, similarly concerned about this new technology, both in how it created new risks to the public but also because of how it might possibly reconfigure established economic norms and power structures, passed legislation that required city streets to have speed bumps every one hundred feet? And then followed that up with additional laws that put hard restrictions on overall vehicle size and weight?

Now let's say that somehow these interventions stuck and for the most part drivers complied with them. Presumably, streets would be safer now. There would also probably be more walkable cities, more public transit, less suburban sprawl, fewer environmental impacts. If you're a cosmopolitan urbanist, what's not to like? If you lean in a more nativist, MAGA direction, you might, surprisingly enough, think the same—because life in America would undoubtedly be slower, smaller, more provincial, more homogeneous.

At this point, cars have been such a central component of American life that it's hard to fully grasp how much they improved standards of living—but, paradoxically, easy to think we can imagine life without them. In the twenty-first century, when you can work from home, when delivery services of various kinds will bring everything to your door, when you can hang out with your friends on Twitch or Discord, what do we really need cars for? And given all the problems they've caused, was it a mistake to embrace them so wholeheartedly? Couldn't we have gotten from there to here without them, albeit at a slower pace?

Not really, no. Today, we have firsthand experience with all the problems cars cause, but less perspective on all the problems they helped

* Most cars at the time were already capable of hitting higher speeds, around 40–45 mph, but speed limits were becoming more commonplace. In 1913, California imposed maximum speed limits of 15 mph in residential districts, 20 mph in business districts, and 35 mph on country roads.

solve. As historian Clay McShane documented in his 1994 book, *Down the Asphalt Path*, early automobile advocates proposed cars as the *solution* to urban traffic jams, accidents, and pollution. As industrialization, urbanization, and commerce had increased in late-nineteenth-century America, horses had become a primary component of how cities operated, particularly for last-mile freight delivery. In 1900, approximately 130,000 horses were pulling wagons and carriages through the streets of Manhattan every day, and leaving approximately 1.3 million pounds of manure in their wake.[21] "Even at the relatively slow speed at which most traveled, horses caused accidents by biting, kicking, and bolting," McShane wrote. Reports from the New York City Board of Health showed that the city's annual death rates from horse accidents ranged between 5 and 6 per 100,000 in the late 1890s. Estimates suggest that as much as one-third of the nation's farmland was needed to fuel horses in that era.

At the dawn of the twentieth century, the limitations of genuine horsepower had already been well understood for decades. As early as 1868, a Paris-based correspondent for the *New York Times* writing about local rumors of "a small steam locomobile for one person for the streets and common roads" noted that "a cheap mechanical substitute for the horse" was an idea whose time had come. And as McShane recounted in his book, various kinds of road-operated omnibuses and other steam vehicles date back to the 1820s and 1830s. By the latter decades of the 1800s, prototypes for carlike "steamers" were increasingly common, but people had deep reservations about sharing urban streets with them. "The pre-1890s steamers failed mostly because of regulation, not mechanical inefficiency," McShane wrote. "They delivered greater speed, lower operating costs, and less pollution than horses; but the public still feared them."[22]

That was understandable, because steamers were a fairly radical addition to nineteenth-century city streets. They were much faster than horses. They left smoke and steam exhaust in their wake. Sometimes they blew up. So, even as technological advances in the 1870s and 1880s led to improvements in performance and safety, local lawmakers consistently banned them from urban areas.

In this same general time period, however, iterative deployment played out in the form of increasingly prevalent bicycles, cable cars, and trolley cars. Traffic was getting faster, heavier, and more diverse. All these changes had impacts on the way city residents perceived and used streets. They'd been seen as a space for socializing, recreation, and transportation, but the focus narrowed as vehicular traffic of all kinds took priority.

By the 1890s, when automobiles with internal combustion engines began to appear, conditions and sentiments had shifted. Not that there was anything like consensus upon their arrival. In fact, there was a great deal of resistance—but not the kind of outright bans that had thwarted the adoption of steamers. Instead, regulations were drafted and deployed in an iterative fashion themselves, as specific risks and harms materialized.

In 1901, Connecticut passed the first speed limit for cars in the U.S.[23] Two years later, a New York City businessman named William Phelps Eno proposed a set of "rules for driving" that the New York City Police Department ended up adopting and enforcing.[24] In 1914, Cleveland installed the first electric traffic light in the U.S.[25]

Ultimately, though, it was unfettered experimentation that informed these interventions and continued to drive progress. Much of this activity was also explicitly incautious in nature. "Speed tests" and car races were so popular in the early days of automobility that Henry Ford, after seeing his first company go bankrupt after just eighteen months in business,[26] recognized that one way to gain the engineering credibility he would need to attract investors for his next venture was by winning a "sweepstakes" race at the local track.[27]

In 1904, the Automobile Association of America started organizing road rallies known as Glidden Tours, with dozens of participants attempting to complete long-distance routes across unpaved wagon roads, with only hand-sketched maps as guidance.[28] In the early 1920s, Studebaker dealers in Southern California promoted annual "economy runs" to demonstrate the brand's reliability, durability, and efficiency under a range of different driving conditions. In a kind of automotive benchmarking, Studebakers that had already been in service for at least 50,000 miles set out from downtown Los Angeles to a mile-high midpoint in the mountains at Lake Arrowhead, then returned to the city. The winner was the vehicle that used the least amount of gasoline, oil, and water to complete the 172-mile journey.[29]

Traveling long distances on poor roads in remote and challenging areas in vehicles that were still evolving technologically meant that danger and sometimes devastating mishaps characterized even the nonracing events. But such risky endeavors also had positive impacts on safety, as the mandate to perform well in races, road rallies, and economy runs led to more reliable and capable cars. Additionally, in what may be the most surprising aspect about the permissionless innovation that helped shape automobility's rise, innovations that emerged from this open and experimental environment weren't only technological in nature.

William Phelps Eno held no public office when he started advocating for his new rules for driving; he was simply acting as an engaged citizen. When automobiles started showing up in New York City in greater numbers and city agencies did little to impose order, Eno drafted a set of recommended practices and persuaded the city's police commissioner that the commissioner did in fact have the authority to adopt and enforce them.

When the commissioner told Eno that his department lacked the budget to print up pamphlets that would be needed to inform the public about the new rules, Eno footed the bill for the first 100,000 of those—and then continued to pay for "all the printing used in traffic work" for the next seven years.[30] Similarly, it wasn't federal regulators leading the charge for greater fuel efficiency at the dawn of the automobile age; it was dealers looking for a sales hook.

In these early years of iterative deployment, resistance persisted, perhaps even increasing as automobility began to gain a real foothold in the culture. Early critics characterized automobiles, aka "devil wagons" and "death cars," as toys for the rich. They tagged drivers as "speed maniacs" and "road hogs" afflicted with "gasoline rabies."

As of 1907, the *Economist* was still rooting for "the triumph of the horse." That same year, in the annual Glidden Tour, someone fired two shots into the body of a participant's car. Even so, the driver told journalists at the end of the race that the locals he'd encountered were actually much friendlier than they'd been in previous years.[31]

In 1909, farmers near Sacramento had "dug ditches across roads . . . and actually trapped thirteen cars."[32] In Europe, where the first internal-combustion cars had been invented, the opposition was more formal. Britain imposed heavy taxes on cars and gasoline. In France, which had briefly outpaced the U.S. in annual automobile production, gendarmes in Paris were instructed to shoot out the tires of drivers who exceeded the speed limit.

Effectively mitigating the harms that cars create is a process that continues to this day, through formal regulations, technological innovations, changing cultural norms, and shifting priorities. In time, though, precisely because it was a public process, and because the proliferation of the Model T and other affordable cars put this new superpower within reach of more and more people, what at first felt alien and threatening to many soon became aspirational, then attainable, then commonplace.

In short, people had an opportunity to choose, and they chose cars. They chose them because cars radically amplified individual agency, in ways that carriages, railroads, bicycles, steamships, and light rail simply could not equal. Essentially cars functioned as new nodes in an increasingly dense and expansive network of streets, roads, and highways that enabled millions of people to collapse time and space in highly individualized ways. As this network grew, it increased optionality for everyone, including those who never laid their hands on a steering wheel. Better transportation networks created bigger markets for all kinds of goods and services. Bigger markets and lower distribution costs led to lower prices, which led to more and even bigger markets. And not just bigger commercial markets, but also bigger and more diverse markets for *everything,* including schools, employment opportunities, places of worship, lifestyles.

Automobility was a flywheel for self-determination and cultural innovation that diffused across society. Permissionless innovation enabled technological breakthroughs, broad cultural adoption and adaptation, and increasing levels of safety. Formal regulation played a role in this process, but it was itself largely iteratively deployed rather than preemptively imposed.

Over time, the results of this approach were stunning. Driving today is not a risk-free proposition—that's a major reason for all the investment that has gone into developing autonomous vehicles. But driving has gotten substantially safer over the years. "In 1923, the first year miles driven was estimated, the motor-vehicle death rate was 18.65 deaths for every 100 million miles driven," the National Safety Council writes on its website. "Since 1923, the mileage death rate has decreased 93% and now stands at 1.33 deaths per 100 million miles driven."[33]

As a general template, the approach we took with automobility makes sense for AI too. Instead of depending on regulators and industry experts to develop and refine AI behind closed doors, in centralized, undemocratic ways, we should continue to engage in iterative deployment that helps us better understand how people are using AI, see where issues develop as usage scales, and adjust accordingly. Through this process, people will get a firsthand sense of how they value, or don't value, the new capabilities that AI affords.

That, in turn, will help determine what kinds of risks and trade-offs seem reasonable. If all that AI delivers for most people is a convenient way to make images for homemade birthday cards, we as a society probably won't tolerate much risk at all. On the other hand, if most people come to see AI as a technology that can amplify their agency and expand their life choices, in the way that automobility has over the last century and a half, then we'll tolerate a higher level of error and risk in pursuit of these greater rewards.

CHAPTER 7

INFORMATIONAL GPS

Until the early 2000s, "paper maps" did not exist—they were simply called "maps." Unwieldy, hard to update, and honestly, if you think texting while driving is not a great idea, try paper-mapping while driving. It was a dangerous era.

In 1973, the Department of Defense began work on what would eventually become the Global Positioning System, or GPS, which uses radio signals from multiple satellites in medium Earth orbit to pinpoint the geographic coordinates of receivers on the ground.[1] By the end of the decade, the U.S. Air Force had begun testing a fledgling version of the system, for military use only.

Then, in 1983, the Soviet military shot down a Korean passenger jet that had flown off course into Soviet airspace. In the hope of averting similar catastrophes, President Ronald Reagan announced that whenever GPS became fully operational, the United States would also make it available for civilian use. Thus, with this grandly humanitarian gesture, a president known for his hawkishness and reducing the size of government paved the way for a free global public utility that has become an indispensable resource for navigating the twenty-first century.[2]

In 1989, the Magellan Corporation introduced the first handheld receiver for the consumer market. In 2023 dollars, the Nav 1000 cost $7,727.02. The sticker shock was made worse when, less than a year later, anyone who bought the device experienced a major downgrade in capabilities when the U.S. Air Force imposed a new policy called "Selective Availability." Out of concerns over national security and other potential misuses, the Air Force deliberately scrambled the signal available for civilian use to make it ten times less accurate than the real thing.

Still, there was enough demand from shipping and logistics companies, surveyors, and other early commercial adopters that Magellan and other manufacturers had begun to generate around $1 billion a year in sales by 1994.[3] Inspired by this growth and the possibility for more, especially after the collapse of the Soviet Union had reduced global security concerns, President Bill Clinton announced in 1996 that the government would soon end Selective Availability and give civilians access to the same level of service the U.S. military enjoyed. A completely accessible GPS, the reasoning went, would boost private-sector investment and innovation, accelerate adoption rates, and dramatically increase the overall value of GPS as a global public good.

At midnight on May 30, 2000, when this new level of access was operationalized, there were only around 4 million civilian GPS users worldwide. But the price for a simple receiver had dropped to around $100, and they were getting smaller and cheaper so fast that they were starting to be built into laptops, watches, and cell phones. A subsequent government policy decision provided a further boost to the market: an FCC requirement that started going into effect in the early 2000s made it mandatory for cellular carriers to provide 911 call centers with the caller's latitude and longitude whenever a caller dialed 911. Carriers fulfilled this obligation primarily by incorporating GPS into phones.

"Devices that know where they are will soon be everywhere. And everything is going to know where it is," James Spohrer, IBM's chief technical officer, told the *New Yorker* in 2000. "We are going to map every meter of this planet."

No doubt that sentiment struck many who read it at the time as a terrifying—or at least distressing—step forward. Twenty-five years later, it may resonate even more strongly now that we have not only mapped nearly every meter on the planet but Yelped them as well. After all, do we really want to turn every otherworldly lava tube in the wilds of Northern Patagonia into terra cognita? Do we reduce the scope of human existence by making the world so legible we completely lose our ability to get lost?

In some ways, yes. In others, no. Knowing every meter of Northern Patagonia makes it easier to track the changes it is undergoing and protect it. Getting lost can be enriching, but recognizing that you're always equipped to reliably navigate toward a target destination is also incredibly empowering. Big Knowledge gives us the power to go on road trips with zero planning and synchronize plans with friends along the way. It helps us discover unheralded restaurants in unfamiliar neighborhoods. It helps us locate a hard-to-find trailhead faster, so we have more time to hike. This capacity to thread logistical needles and venture into the known unknown makes life richer. In shrinking the physical world to fit into the palm of our hands, we expand our horizons.

What happens, on a collective level, when everyone suddenly has new navigation and coordination superpowers? A 2019 report from the National Institute of Standards and Technology estimates that GPS technologies created $1.4 trillion in economic benefits to the public sector from 1984 to 2017, with 90 percent of that accruing in the last seven years of the period.[4]

While turn-by-turn navigation may be the most common way we benefit from GPS, it's far from the only one. The precise timing information GPS provides is used to synchronize clocks in telecommunications networks, in ways that help keep mobile phone calls clear and lag-free. During natural disasters and other emergencies, first responders use GPS-enabled drones to locate missing persons, quickly map stricken areas, and even deliver supplies to those who cannot be easily reached. Precision-farming techniques that GPS enables make organic arugula more affordable. Not bad for a technology whose chief architect, U.S. Air Force colonel Bradford Parkinson, originally described it as an attempt to "drop 5 bombs in the same hole."[5]

New Tools for the Road Ahead

What does this extended detour have to do with AI? First, it maps out a clear example of the positive outcomes that can result when the government embraces a pro-technology, pro-innovation perspective and views private-sector entrepreneurship as a strategic asset for achieving public good.

Second, it's also a great example of how we can effectively leverage our capacity to turn Big Data like geographic coordinates and time stamps into Big Knowledge that can be used to provide context-aware guidance in many aspects of our lives.

Third, and most importantly for democracy, it reinforces individual agency.

While GPS serves many purposes, across multiple domains, its breakthrough application was turn-by-turn navigation. It and the commercial services built on top of it allow us to move through the physical world with constantly updated knowledge. At literally every turn, these navigation systems increase individual agency by telling us where we are, what else is nearby, what obstacles we need to dodge, and so much more.

LLMs and the conversational agents built on top of them function similarly: they increase our capacity to navigate the complex and ever-expanding informational environments that define life in the twenty-first century. In doing so, they're now enhancing the individual agency of people worldwide, by providing the kind of situational fluency that lets us make better-informed decisions and get ourselves where we want to go.

This is especially important given the centralizing tendencies of AI, which requires extensive data, hardware, energy, and human talent to achieve state-of-the-art performance. To continue to develop AI in alignment with the democratic ideals of self-determination and broad participation, we need to design and roll out AI in a manner that prioritizes individual agency and gives people hands-on access to tools that they can use in practical, open-ended ways. Conceptualizing LLMs as a form of informational GPS provides a familiar model.

Along with their similarities, there are clear differences between GPS and LLMs. In the case of the former, the U.S. military exercised exclusive control over the development of the core technology. The latter are the product of a global array of academic researchers, large corporations, open-source advocates, and startups, and are available as open-source, proprietary, or partially open models that offer limited developer access through controlled access points known as application program interfaces, or APIs.

Even more fundamentally, GPS deals primarily with objective, ground-truth spatial and temporal data—that is, geographic coordinates and precise time stamps. LLMs create outputs based on the nuances, complexities, and subjectivity of human language. While LLMs can process factual information, there is no single overarching canon of objective information and "truth" for them to utilize, especially when dealing with the more widely interpretable aspects of human knowledge.

Instead, every LLM developer effectively creates a unique "informational planet" of its own making, along with a unique map of that planet. In other words, no matter how large that training datasets get, they'll never contain all possible information. In addition, every developer effectively maps the planets they create in a unique way. Two different developers might start with the same exact dataset, and yet because they employ different numbers of parameters and weights, different algorithms, and different fine-tuning techniques designed to align the model with specific preferences and values (which will vary from developer to developer), informational planets are emphatically human constructions in a way that physical planets are not.

While the earth largely stays fixed, these informational planets can change over time, through updates in training techniques and the incorporation of new data. Business models can potentially have an impact too. Imagine, for example, how developers might filter training data or tweak optimization algorithms in an attempt to overprioritize content that drives engagement and ad revenue at the expense of accuracy and representativeness. Finally, since LLMs generate outputs based on statistical probability rather than fixed rules, a single prompt—or "request for directions," to draw upon the GPS analogy—can produce different outcomes each time you input it.

Nevertheless, there are important ways in which GPS remains a useful analogy for understanding the role that AI will have in our lives. Consider how we are not only constantly navigating the physical world, but also a complex informational world. There are beliefs, values, and traditions that contribute to your identity as a member of a particular community or group; specific knowledge and lexicons you use in your line of work; laws, rules, and norms that define your life as a citizen; additional literacies that cut across the culture. How do you distinguish between fact and opinion in each of these different cognitive territories? What conventional wisdom and what counternarratives inform them? What shared knowledge and modes of discourse are you expected to master to qualify as a local in each distinct place?

Life as a human today means constantly upskilling—at work, yes, but everywhere else too. While ongoing digital innovations perpetuate this dynamic, they also help us manage it. To keep up with new informational demands, the twentieth century brought us new products and services like email, hyperlinks, search, and emojis. The twenty-first century has given us AI.

As their name implies, large language models are, at heart, systems for analyzing, synthesizing, and mapping language flows. That's what informs the analogy to GPS navigation systems—LLMs are infinitely applicable and extensible maps that help you get from point A to point B with greater certainty and efficiency. You can ask ChatGPT to "translate" a white paper on Q-learning in terms that will help a person with no computer science background form a basic understanding of the material it covers. You can have it evaluate the terms and conditions for a proposed contract with a potential client in a country where you've never done business and have no knowledge of standard market mores.

Offering accelerated fluency and navigational support on demand, LLMs become keys to heightened participation and competence in a world constructed from information. As such, they're a democratizing force, echoing trends we've seen in earlier instances of technological innovation. Thanks to blogs and social media, for example, millions of people now regularly perform functions that were once the near-exclusive domain of professional journalists. Ride-hailing services like Uber and Lyft have similarly turned an industry that was strategically restricted through expensive barriers to entry into one where millions can easily participate. With AI, people who once lacked the skills, training, or resources to produce different kinds of written materials, illustrations, computer code, music, and video can now do so. As models evolve, they also become more adept at emulating the work of lawyers, tutors, therapists, and other professionals.

Because LLMs are so broadly applicable and extensible, they may conceivably compete with virtually all of us in *some* way. When it's your livelihood bearing the brunt of such democratization, the shock of this new reality can obviously be disorienting and distressing. But when we orient LLMs around individual hands-on use, we can apply their extraordinary versatility to our own ends in *many* ways. This approach helps us continue the dominant trend that has characterized computerization, automation, and digitization since the 1960s—an increase in individual agency and autonomy through tools that enable greater creative expression, productivity, and impact. The more adept you become at using LLMs to navigate life in the twenty-first century, the greater your power to plot your own path through the world. And what's more democratic and dynamic in the long run: societies that strive to increase choice and autonomy for everyone, or ones that protect existing conditions for entrenched incumbents forever?

License to Skill

Mobility has always served as the foundation for self-improvement. In the 1700s, the increasing accessibility of commercial transoceanic travel made it possible for a young man born into a farming community in rural England to pursue an entirely different fate as a merchant in Philadelphia; geography was no longer destiny. The growth of railroads in the 1800s meant that a seamstress living in rural Ohio had increasing access to more kinds of textiles and patterns and could potentially sell the clothes she produced to a much larger market in Chicago. In the twentieth century, a car enabled a twenty-five-year-old family man to find a job where wages were highest and a home where mortgages were most affordable.

In more recent decades, the internet and mobile phones took the place of the flying cars we were promised. Press a few buttons, and we can transport ourselves through space via apps like Zoom and Google Street View. Press a few more, and everything from Whole Foods sweet chili salmon to a 600-watt portable power station arrives at our doorsteps, almost as quickly as if we'd created them with a Star Trek replicator. Just as easily, we can retrieve facts, connect with others, and otherwise experience and understand informational and social landscapes with hypermobile facility.

In the same way it's vital now to have GPS if you're visiting a foreign country for the first time, don't speak the language, and are trying to make sense of wildly unfamiliar terrain, informational GPS can give anyone who uses it more situational awareness, in almost any context. A number of studies published in the first half of 2023 showed how incorporating AI into workflows increased productivity across a range of industries. Of particular note: many of these studies also showed that gains were highest among less-experienced participants.

In March 2023, for example, MIT researchers asked 444 college-educated marketers, grant writers, data analysts, and other professionals to perform short writing-related tasks, like composing press releases or preparing analysis. According to their findings, workers using Chat-

GPT completed tasks 37 percent faster than those who didn't use it.[6] That April, another study showed that customer service agents at an unidentified company increased their productivity by 14 percent when using ChatGPT as a real-time assistant during their customer interactions.[7]

In the study involving the college-educated professionals performing writing tasks, "the quality boost was bigger for participants who received a low score on their first task," which they'd been instructed to complete on their own, without ChatGPT's assistance. In the customer-service study, the employees who saw the biggest gains were "less experienced, lower-skilled customer service reps."

This makes intuitive sense. If you've lived in, say, Tokyo your entire life, GPS navigation will likely still be helpful in some instances. But you probably won't find it nearly as useful as a first-time visitor would, especially if that visitor has no fluency in Japanese or much experience navigating large cities.

For LLMs too, a similar dynamic is in effect. Having been trained on so much data, LLMs are generally well acquainted with what might be described as standards of competence in a wide range of contexts and knowledge domains—because that's what is represented most frequently in their training data. For example, most models will quickly recognize that 2 + 2 = 4, because there are a lot more instances of text strings on the web that say "2 + 2 = 4" than there are ones that say "2 + 2 = burrito." You might say that, even with their extraordinary capacity to synthesize information in nuanced ways ("Explain climate change using concepts from classic music theory"), an LLM's "natural instinct" is to think in clichés. So, an AI is generally quite adept at recognizing what constitutes a competent customer service interaction for a given situation, or what should be conveyed in a press release promoting a special limited edition of Doritos.

In this way, LLMs can help upskill beginners very quickly. "It has almost like a democratizing effect," economist Robert Seamans of New York University's Stern School of Business told *Nature*.[8] "The workers who are less experienced are the ones who would benefit more from it."

Because of its inherent scalability and adaptability, machine intelligence has an additional democratizing effect. While knowledge is broadly accessible in the form of books, videos, and other forms of media, human intelligence tends to be more tightly clustered, given how it's bound up in actual human brains. There are far more psychiatrists per capita in cities like New York and Seattle than there are in rural counties. Legal experts cluster at high-priced law firms; computer science PhDs with expertise in machine learning gravitate toward Big Tech campuses.

While LLMs do not possess intelligence in the same way that humans do, interacting with them already provides a much different experience than interacting with a Wikipedia page or a podcast. This will become even more apparent as LLMs acquire more multimodal capabilities. As the technologies evolve, it is becoming increasingly possible to receive information, highly personalized to your needs, in whatever media format you prefer.

Ultimately, people across all strata of society will benefit from this technology, just as they do from GPS and smartphones. But as synthetic intelligence diffuses throughout society, it's likely to have an especially transformative impact on those who lack access to the places where human intelligence traditionally clusters.

For example, if you're a high school student from a low- or middle-income family navigating the college application process with little access to pricey human tutors, coaches, and advisors, LLMs may offer context and guidance that helps increase your chances of admission.

If you get a letter from your landlord that says you must "quit premises" in thirty days unless you pay $3,000 in "rent arrearage," an LLM can offer suggestions about your potential next steps. If you're a non-native speaker who struggles with English, let alone unfamiliar legalese like "rent arrearage," you could use your phone's camera to share the letter with an LLM, which could then explain to you, in your own language, what the letter means and what your options are. In a more forward-looking scenario, a reliable AI assistant might even volunteer to contact local legal aid services on your behalf to see what resources might be available.

An LLM's ability to "translate" from one medium to another has many similar applications. A student with dyslexia could use a multimodal LLM to convert text-based lessons and other learning materials into different kinds of audio formats. (For example, a tool called Illuminate that Google Labs demoed at the company's annual developers conference in May 2024 converts PDFs of highly technical research papers into accessible podcast-style conversations geared toward general audiences.)[9]

Multimodal LLMs will also further advance the kinds of capabilities found in existing services that help people with vision impairments interpret the world around them. Such a user could scan items with his phone while shopping and have a detailed conversation with the LLM about different products, asking about prices, what kinds of warnings or disclaimers are on a label, whether or not a competing product is on sale, and so forth.

A deaf person could use multimodal LLMs in similar ways. The AI could provide real-time, highly accurate transcriptions of what another person, or persons, are saying to them, along with contextual cues about the speaker's tone of voice ("he said sarcastically"), speaker IDs (if more than one other person is involved in a conversation), and more.

Likewise, the LLM could translate the deaf individual's signing into audio responses for the benefit of hearing individuals who don't know sign language. LLMs could also provide increasingly sophisticated and contextually aware sound-recognition capabilities, improving on existing apps that notify hearing-impaired users about important environmental sounds, such as alarms or approaching vehicles.

Even more than search or Wikipedia, LLMs can provide clear and easy-to-access starting points for information-gathering. Instead of typing queries into Google and then trying to evaluate which links are genuinely helpful, you can just start having a conversation with an immediately responsive and informed guide.

A feature installed on Google Pixel 9 smartphones, called Screenshots, suggests another way that multimodal AI will become useful to everyone, especially to seniors. Essentially, it makes photographic memory universal. When you take screenshots of information you want to keep track of—a text from a friend, that time you got Wordle in one try, the address of a medical lab you're going to need to return to in a few months—Google's AI processes the image as a text-based record. Then, whenever you want to return to it, you can retrieve it instantly simply by saying "medical lab address" or "that Wordle hole-in-one I got a while back."

When technology solutions are presented as ways to make education, health care, and other essential services more accessible to underserved communities, they're often dismissed as Band-Aid fixes to much more deeply rooted structural inequities. "The rich get their concierge doctors, their private trainers, their $400-an-hour therapists," this line of thinking goes. "Everyone else gets a complimentary 'wellness' app from Blue Cross and a dog-eared copy of *Personal Bankruptcy Laws for Dummies!*"

In other words, the presumption is that complex and high-value services like legal counsel and health care can *never* be effectively automated and scaled, because the results won't ever be anywhere near the quality that high-level human practitioners can dispense. Consequently, such efforts are often written off as misguided altruism, or tech industry grifts, or simply solutionism theater deployed with no genuine intent to deliver value to the recipients.

But the growing capabilities of LLMs are starting to challenge this narrative. While we believe that human professionals working in tandem with LLMs in domains like education, health care, and the law represents the ideal scenario in most cases, LLMs on their own, working directly with a user, can provide great value. They aren't error-free, but neither are human practitioners.

Unlike their human counterparts, LLMs are instantly accessible, infinitely patient, and always willing to answer just one more question. If you leave them hanging at 3 p.m. only to return at 1 a.m. they will still remember exactly what you were discussing, in detail. Ghost them as many times as life necessitates: they never hold a grudge. For a busy executive, this is a nice perk. For a low-income, underresourced person who may have limited transportation options, an inflexible work schedule, and scarce childcare options, this kind of accessibility could be life-changing.

In other words, we're entering an era when intelligence will no longer be a limited resource. Automated and networked, it will meet people wherever they are and become a ubiquitous and indispensable part of daily life for billions of people, as GPS now is. Just as we reflexively turn to our smartphones for directions in unfamiliar territory, or rely on GPS for precise time synchronization in financial transactions, we'll utilize AI in ways that fundamentally reshape problem-solving and decision-making in both personal and professional spheres. Everyone will benefit from this shift, but for people whose ability to effectively navigate informational landscapes has been constrained for various reasons, the new capabilities and freedom that AI brings to their lives will be especially profound.

Now You're Speaking My Language

As we've noted earlier in this book, developers are starting to build LLM-based systems capable of acting more agentically themselves. For example, OpenAI's GPT-4 with Code Interpreter can autonomously write, run, and debug code, to execute even complex programming tasks without step-by-step user intervention. AutoGPT is an open-source application that uses GPT-4 to execute a series of tasks designed to accomplish a specific goal, such as creating a recurring summary of new research papers on a given topic that are published each week at arXiv.org, an open-access repository where scientists often share their work before more formal publication.

Although they typically operate within carefully defined constraints and allow for human intervention and oversight, agentic AI systems go beyond the simple question-answering that is the forte of traditional chatbots. The objective is to use LLMs in ways that enable long-term planning and self-improvement through learning. In this way, a more agentic system can engage in more humanlike workflows and function more like an assistant or coworker.

Imagine an AI that can help you identify up-and-coming influencers who are a good fit to promote your new fitness app. It would start by drafting an initial set of steps to pursue, then it might write code to automate data collection from social media platforms or perform sentiment analysis on the comments and reviews posted by a number of influencers. As it evaluates information, your AI agent might refine its tactics if it's not finding influencers with high enough engagement rates in the fitness niche. In some systems, multiple agents could be incorporated to work on specific aspects of the project, much like a team of human colleagues working collaboratively to fulfill a shared goal.

Many people believe more agentic LLMs will completely supersede "LLMs that just talk to you." That's because agentic systems will enable synthetic intelligence at scale in a way that traditional conversational AI can't. Agentic AIs will theoretically work on multiple complex tasks simultaneously, with minimal human oversight. So, it's the difference between doing ten things, or a hundred, or a thousand things at once, compared to doing just one task at a time. This ability to multiply human productivity and tackle numerous high-value tasks in parallel makes agentic LLMs particularly attractive for businesses and individuals seeking to maximize their impact.

And yet even as more agentic forms of AI grow increasingly popular, we believe engaging in multi-turn dialogue with LLMs will remain a key part of maximizing their value. As LLMs and the systems built on them grow capable enough to operate autonomously in highly reliable and adaptive ways, these advances will also make them better at listening, interacting, and following instructions in ongoing one-to-one conversations with them. As that happens, you're going to be able to do more, with easier, more precise control, and you're going to want to take advantage of these systems as much as you can.

Much of what initially made ChatGPT so transformational involved how it explicitly required users to play an ongoing role in the process of using AI. In this mode of interaction, an LLM literally does nothing until you enter a prompt. Once the LLM replies to your prompt, it does nothing again, unless you prompt it further.

This turn-based interaction with LLMs requires a high level of engagement—and that obviously helps mitigate some of their weaknesses. By reviewing every output that you get from an LLM, you're more likely to notice if it hallucinates or generates some other kind of incorrect information. (This, at least, is the optimistic view. The Gloomer take is that you're presumably more likely to interpret its confident assertions as truth.) Similarly, if it produces biased, toxic, or other undesirable outputs, you can challenge or correct it.

This openness to instant user feedback immediately distinguished hands-on LLMs like ChatGPT from most earlier forms of AI. Structural biases in AI models are an important issue, and the problem tends to be exacerbated by the fact that the individuals most affected by such bias have generally had no way to interact with the model and counteract the harm. In the case of algorithmic decision-making in loan approvals, for example, a person who is impacted by a discriminatory algorithm may not even have any idea that its decision was the basis for their rejection.

With conversational AI, it's different. Some LLMs even allow you to prime them before you ever issue a prompt, by creating custom instructions that convey your values, intents, and the kinds of responses you're looking for. Granted, this functionality isn't foolproof. If you tell ChatGPT, "Don't hallucinate!" it may still hallucinate. If you say, "Write so beautifully it will make statues weep and win me a Pulitzer!" that's not going to happen. The AI is just not ready yet, even if you command it to be.

But the ability to intervene, correct, and fine-tune LLM outputs in real time affords them a capacity to meet you where you are—in a way that books, video, and other forms of mediated knowledge simply cannot. If you prompt ChatGPT to "Create an image of a flight attendant serving lunch to a group of lawyers," it will probably depict the flight attendant as female and the lawyers as predominantly male.* But if you want a different image, you can steer toward the result you desire. You might say, "Create an image of a male flight attendant and a female flight attendant serving lunch to a gender-balanced and ethnically diverse group of lawyers."

This capability doesn't magically negate the underlying issue of structural bias within a model's training data. Making models more representative and inclusive is an ongoing process. But, more than with other kinds of media and other kinds of AI, conversational AIs give users unprecedented opportunities to exert their influence. You can guide, question, inform, and push back against the AI, and thus redirect the content and flow of a discussion.

Your ability to intervene in constant and powerful ways, paired with the near-cosmic cultural omniscience that LLMs acquire from mapping galaxies of data in obsessively precise ways, allows you to nudge and tweak their outputs like a recording studio engineer manipulating faders, knobs, and switches on a 128-track mixing console. This structural capacity for specificity can be applied toward mitigating bias—and toward countless other needs and goals. Whatever your particular emotional, cultural, professional, ideological, or philosophical language, an LLM can likely speak it (albeit with varying degrees of fluency).

With some caveats, that is. Most developers employ a variety of techniques during a model's fine-tuning to establish safety guardrails and policy norms that may constrain responses to certain kinds of prompts that violate the core principles and ethical guidelines that a developer has set. But these modest limitations leave users tremendous

* This was true when we were writing this chapter. But given that developers are constantly updating models in their attempts to reduce issues like bias, it's possible that, by the time you read this book, the same prompt will produce different outputs.

latitude to shape an LLM's responses to their unique goals and preferences.

Think of Google Maps: zooming in and out on a given locale is just the start. You can also view it through different layers—one that shows you topography, one that reveals live traffic, one that indicates air quality, one that shows you bike routes. As you cycle through layers, zoom in and out, and shift to Street View, you see new angles and details, new facets of the landscape. Similarly, turn-based dialogue with conversational LLMs can help you "map" the factual, emotional, political, cultural, economic, and historical terrain of subjects and situations, in whatever ways are most relevant to your current needs and intentions.

A simple way to illustrate this is to try a series of prompts such as:

"Explain the theory of relativity to a six-year-old."
"Explain the theory of relativity to a high school physics student."
"Explain the theory of relativity to an adult layperson."

Then ask the LLM to process this request using even more granular instructions, to tailor the answer for any of an infinite number of six-year-olds:

"Explain the theory of relativity to a six-year-old who loves fire engines."
"Explain the theory of relativity to a six-year-old whose primary language is Spanish but who is also in the process of learning English."
"Explain the theory of relativity to a six-year-old who is fascinated by outer space and planets, and prefers learning through storytelling."

In this respect, LLMs both accelerate and reward fluency in a given domain. The more knowledge and nuance you can apply, the better you can tap what Ethan Mollick, a professor at the Wharton School who specializes in entrepreneurship and innovation, has described as the "latent expertise" of LLMs—which is to say, all the knowledge they've absorbed implicitly through their training, in ways that are neither immediately apparent nor entirely predictable. "Experts thus have many advantages," Mollick writes in an essay at his website, One Useful Thing. "They are better able to see through LLM errors and hallucinations; they are better judges of AI output in their area of interest; they are better able to instruct the AI to do the required job; and they have the opportunity for more trial and error. That lets them unlock the latent expertise within LLMs in ways that others could not."[10]

The power of expert instruction was on vivid display in the early days after OpenAI released DALL-E 2, its text-to-image generator, in the spring of 2022. Photographers, illustrators, and others who were highly fluent in the grammar and lexicon of image creation were at the forefront of unlocking DALL-E's latent expertise through the specificity of their prompts. While many early adopters were relying on broad prompts to generate images ("an astronaut riding a horse on Mars!"), visual professionals were instructing it to mimic specific camera lenses and light sources, and adding compositional directions ("an astronaut riding a horse on Mars, 50mm lens at f/2.8, illuminated by the warm glow of the setting sun reflecting off Phobos, with the subject placed off-center"). Expert input from human users was what began to uncover the full range of capabilities that DALL-E 2 could draw upon.

Wherever your own particular domain knowledge lies, the concept of "latent expertise" has implications for you, as it does for all LLM users. In any context, specificity matters, because specificity is how you can best tap into how any given LLM has mapped its own informational planet. And even if you're a newcomer to the domain you're trying to crack, completely unversed in its lexicons, norms, and nuances, you're an expert when it comes to knowing what you're trying to do right then and where you think you want to go. That gives you the point of entry you need to start unlocking an LLM's latent expertise.

When you're interacting with LLMs, it's useful to provide as many "coordinates" as you can: What are you seeking to learn? ("I want to understand the basics of quantum computing.") Is there a specific goal or intent behind that request? ("I'm preparing for a job interview at a tech company.") What details about *you* might help the LLM tailor a response? ("I have a background in classical physics but no experience with quantum mechanics.") Is there a specific role or persona the LLM itself should assume for this interaction? ("Please explain this as if you were a patient high school science teacher.") What factors might make its outputs more relevant to you? ("I learn best through analogies to everyday objects.")

The more you can tell an LLM about where you are and where you want to go, the more precisely it can help you plot a path to get there.

A Step in the Right Direction

As we noted at the start of the chapter, not everyone agreed that releasing GPS access to the public was a great idea when Ronald Reagan first proposed it in 1983. Why make it easy for terrorists, enemy countries, stalkers, and criminals to access such precise geolocation information?

Concerns over what could possibly go wrong led to the Selective Availability policy, which deliberately inhibited the utility of GPS for several years.

In the decades since that policy was repealed, however, we have not seen the kinds of serious threats to national security that the Department of Defense had feared might come to pass with free global access to GPS. That doesn't mean the system is failsafe or risk-proof. In fact, a range of mishaps and abuses regularly occur. The rare but tragic instances where drivers follow erroneous navigation system instructions into remote terrain, or into a body of water, with fatal consequences, draw the most attention. But most people who rely on GPS navigation have experienced similar but generally harmless scenarios at one time or another, with systems providing confusing directions, sending them down private roads or paths, or otherwise leading them astray.

There are also instances where bad actors jam or spoof the precise timing signals that GPS relies on, causing receivers to calculate incorrect positions or fail altogether. Thieves who steal entire shipping containers use jammers to disable the GPS tracking tags on the goods inside them. For the last twenty-plus years, however, the prevailing story of GPS is not a tale of limitations, flaws, and vulnerabilities leading to disruptions in security, navigation, and other essential services. Instead, GPS has globally enabled a wide range of massively beneficial services, every minute of every day, week after week, year after year.

Shouldn't we pursue that same kind of value with broadly distributed, hands-on AI, sooner rather than later? When Bill Clinton announced in 1996 that the government had decided to end Selective Availability, the target date for implementing the switch-over was 2006. But once the announcement had been made, public demand for the higher level of service was so great that the government ended the policy in 2000, six years ahead of schedule. It was a moment when the federal government explicitly chose to accelerate technological innovation and adoption, and it had a tremendous positive impact. We should keep this example firmly in mind as we aspire to make informational GPS more accessible too.

CHAPTER 8

LAW IS CODE

Just a few years after companies like Magellan and Garmin began to deploy GPS to map every meter of the world, Google embarked on a similar quest to map every hyperlink of Earth's fastest-growing and least-charted frontier. As the company declared in its 1998 mission statement, its intent was "to organize the world's information and make it universally accessible and useful."

This was a quixotic undertaking, especially at a time when Google's entire staff could still fit comfortably in a rented suburban garage. But at least it was in the realm of the thinkable. In contrast, the idea that cyberspace might not just be accessible, but also truly governable, was still heresy in 1998. The internet, after all, had been specifically designed to withstand single points of failure—a fact that also made it highly resistant to single points of authority. Operationalized through technical protocols rather than state-imposed laws, it favored network resilience over central control, open access over institutional gatekeeping. "The Net interprets censorship as damage and routes around it," open-source pioneer and Electronic Frontier Foundation cofounder John Gilmore famously said.

In 1999, however, Harvard law professor Lawrence Lessig offered a much different perspective in his book *Code, and Other Laws of Cyberspace*. In it he described how the commercialization of the internet, and the imperatives for more substantive user authentication that commerce demanded, were already changing the internet from a place where unrestricted access, anonymity, and autonomy were the norms to a place that would increasingly prioritize identity, centralization, and architectures of control.

In the real world, Lessig explained, four distinct constraints regulate human behavior: laws, norms, markets, and architecture. On the internet, it was the same, except that in the medium's early years, architecture, in the form of code, played an outsized role. Software developers were constructing this new world from scratch, with little oversight from real-world regulators. So they determined the rules of engagement. On the internet, code was law.

"This code, or architecture, sets the terms on which life in cyberspace is experienced," Lessig wrote.[1] "It determines how easy it is to protect privacy, or how easy it is to censor speech. . . . In a host of ways that one cannot begin to see unless one begins to understand the nature of this code, the code of cyberspace regulates."

What Lessig recognized was that online commerce, and its fundamental need for information, in the form of credit card numbers, mailing addresses, and other personal identifiers, was a kind of Trojan horse, the thing that would necessitate additions to the internet's original architecture. The technological conditions that had once enabled a new level of public privacy, where you could be anyone, anywhere, and say anything without fear of censure, were being supplemented with user IDs, tracking cookies, and the cumulative data collection that these technologies facilitated. "The invisible hand, through commerce, is constructing an architecture that perfects control—an architecture that makes possible highly efficient regulation," Lessig wrote.[2]

There's another way to characterize the impacts of this shift, though. In the internet's early glory days, you had unprecedented freedom to do two things. First, you could interact with fellow travelers on the astral plane in ways that were completely untethered from physical presence, traditional social norms, and any vestige of real-world identity. Second, you could consume the sometimes compelling and often mundane content that other early adopters were sharing with the world, or post such content yourself.

In an internet supplemented by user IDs, tracking cookies, and highly efficient regulation, your freedoms are more diverse. You can, for example, reconnect with your best friend from kindergarten. You can short meme stocks in your pajamas. You can watch more Nicolas Cage movies than pre-streaming Hollywood would have ever thought to produce. You can crowdfund your first robot lawn-mower prototype. You can monitor your grandma's heart rate remotely. And you can take free online courses from Harvard professors, just to name a handful of the twenty-first-century internet's many wonders.

Not that Lessig himself was opposed to such things. As circumspect as he was about the way the internet had begun to change in the late 1990s, he recognized that some regulation was inevitable and, if judiciously applied, even useful. In large part, his objective was to focus attention on the trade-offs that would come with change, their long-term implications, and the fact that this was already a political process in addition to a technical one. Companies, the government, and users were all making choices—or failing to make choices—about how the internet should be constituted, and those choices would ultimately shape the nature of online experience.

That's why *Code* was so important at the turn of the millennium, and why, twenty-five years later, it is even more relevant. This is especially true for its central thesis, the idea that code is law. When Lessig wrote *Code*, the internet, aka cyberspace, was not yet inextricably woven into the "real world." Instead it was a place you entered from time to time through the gateway of your desktop PC or laptop. Outside of cyberspace, the law was still the law.

Today, cyberspace is no more a world apart from the world than, say, telephone lines are or Walmart is. The internet is everywhere now, diffused into everything. Code has migrated into phones, cars, appliances, city infrastructure, manufacturing plants, automated farming systems, health and fitness trackers, implantables and prosthetics, money, and more. Intelligent and perceptive devices permeate our built infrastructure. Many of these new devices have, or will have, the capacity to act in increasingly agentic ways. They'll be making decisions and taking actions of their own volition, in ways that may impact the choices and obligations of humans they interact with.

To better understand the implications of this, let's imagine that it's 2027. While autonomous robotaxis are becoming more prevalent, traditional cars requiring a human driver still dominate new vehicle sales. You've just purchased a Chevy Equinox EV, and like all new cars sold in the U.S. that year, it comes equipped with a federally mandated Driver Alcohol Detection System for Safety (DADSS).

Such systems use touch- and breath-based sensors embedded in various parts of a car's interior to passively measure a driver's blood-alcohol level. Unlike with a Breathalyzer, it isn't necessary to actively blow into these sensors. Instead, they use infrared light to passively monitor drivers. Touch the steering wheel or ignition switch, and sensors embedded there will detect the blood-alcohol concentration levels in the capillaries of your fingers. Simply exhale, and sensors embedded in the dashboard or driver's side door will detect the blood-alcohol concentration levels in your breath.[3]

These sensors don't incorporate AI themselves. But your car is also equipped with more features—a driver-facing camera and additional sensors—that do use machine learning to analyze things like posture, grip patterns, and airflow directions to help determine that it's the driver, rather than any passengers who might also be in the car, whose blood-alcohol content is being measured.

So imagine a scenario in which, after a wonderful dinner that included more than a glass (or three) of wine, you're now arguing with NaviTar, the LLM-enabled interface for your car's entire vehicle-management system, regarding your fitness to drive. NaviTar has determined that your blood-alcohol level is .10, or .02 over the legal limit in California, and consequently, it won't enable the car to start.

You try to explain that the waiter spilled wine on the table and even flash your slightly stained cuff at the driver-facing camera. NaviTar replies that it would be happy to call you an Uber, or play you a movie to entertain you while you sober up. "How about *Speed*?" it suggests. "It's a classic action thriller with just the right combination of suspense, high-stakes drama, and humor to make the time fly by. And it lasts seventy-six minutes. By then, you'll be sober."

"Very funny, NaviTar. And no, no movies," you sigh. "But tell me again. How did we get here? I mean, this is America, isn't it? Don't I have a constitutional right to choose whether or not I want to risk a DUI?"

The technologies that would make this scenario possible all currently exist. In fact, by 2027, such alcohol-detection systems (not necessarily with the LLM component) may be mandatory for all new vehicles purchased in the U.S. In 2021, Congress passed legislation to that effect as part of the Infrastructure Investment and Jobs Act. And while the exact timing of when this will happen was not yet clear when this book went to press, there's a possibility it may be as early as 2026.*

So how does this scenario strike you? Dystopian? Utopian? Maybe a little bit of both? Certainly, as such technologies and scenarios get closer to reality and mainstream use, civil liberties advocates, hospitality industry trade organizations, automobile manufacturers, consumer advocates, and technology skeptics may oppose and perhaps prevent or delay this regulation. Surely at least some of the millions of drivers who routinely ignore seat belt laws and speed limits are likely to resist the proposition that their own cars, so long a symbol of individual autonomy, will effectively turn into traffic cops.

We've talked a lot in this book about the virtues of permissionless innovation, iterative deployment, and the benefit of tolerating some degree of intelligent risk in pursuit of progress, instead of demanding zero uncertainty before introducing new technologies that will change how the world works. But drunk driving is a well-established risk that consistently exacts significant personal and societal costs. Given that Congress included the DADDS provision in the Infrastructure Act, it's obviously a risk our legislators believe we should collectively discourage.

And maybe you do too, even if that means you might find yourself watching *Speed* in a restaurant parking lot at some future date. But others no doubt feel differently. And it's also just one particularly high-stakes scenario. Many other scenarios involving machine agency are just around an unseen bend in the road, hurtling toward us at the speed of the twenty-first century.

* The provision gives the federal government until November 2024 to create a rule guiding how impaired-driving prevention systems will be implemented. The National Highway Traffic Safety Administration (NHTSA) is responsible for evaluating and setting the standard for these technologies. If NHTSA determines that alcohol-detection systems are reliable enough, these systems could be mandated in all new cars as early as 2026.

Who (or What) Is in Charge Here?

At work, your employer might equip your PC with an AI that can sync with your calendar and project management tools to know when you have looming deadlines. If you're falling behind, the AI might put you into "focus mode," disabling all nonessential apps and notifications until you make more progress, and automatically sending a notification to HR.

Your home insurance provider could offer dynamically adaptive systems that use smart sensors to detect leaks, electrical faults, signs of pest infestation, humidity levels in the walls, and more. While you'd get incentives to perform maintenance based on the analytics these sensors generate, your system might, in some "mission-critical" scenarios, deploy more forceful measures to accelerate your compliance, such as automatically decommissioning the ancient furnace you didn't get around to replacing last month. Now it's ten degrees outside, your system has transitioned the furnace to eternal sleep mode, but you can't get an HVAC technician to come out for at least three days . . .

What happens if many employers start making "focus mode" a mandatory condition of employment? What if the home insurance approach described above works so well for providers that they all stop offering other options?

When Lawrence Lessig wrote about code's capacity for "perfect control," this was the phenomenon he was getting at. With other kinds of constraints, including traditional law, individual agency plays a key role. As Lessig expressed it, "*Law* is a command backed up by the threat of a sanction"[4] (emphasis in original).

Municipal street signs, one of the oldest forms of regulatory automation, offer a clear illustration of how this works. NO PARKING AT ANY TIME, a red-and-white sign boldly declares. UNAUTHORIZED VEHICLES WILL BE TOWED AT THE OWNER'S EXPENSE.

For all its unwavering authority, this sign leaves *you* with a decision to make about how strictly to comply. Because in many instances, you probably could park there. Doing so might inconvenience or even endanger your fellow citizens (if, say, you're parking in front of a fire hydrant). But if you truly *couldn't* park there, there'd be no need for the sign.

In this way, almost all laws are largely dependent on your voluntary decision to comply with them. Whatever role professional law enforcement plays in maintaining order and stability, it's just part of a much larger social fabric woven from community cooperation, mutual trust, and widely held norms of behavior—because police officers can't be everywhere. So you might slow-roll through a red light at 4 a.m. with no pedestrians or other motorists in sight. You might not bother to fasten your seat belt when you're just driving a few blocks to the grocery store. You might skip feeding a meter because you will be in and out of that convenience store in less than a minute.

In all these instances, you're making a choice. But a sensorized parking meter, connected to a centralized parking management system, could automate you out of the equation. Park after 6 p.m. in a spot that does not allow that, and if you remain there for more than five minutes, the meter might simply deduct a predetermined fine from your EZ-Park account. In this scenario, a device observes the infraction, enforces the law, and implements the punishment.

At some point, between ubiquitous cameras and sensors, and code-driven perfect control, enforcement could potentially get too cheap to meter. Traditionally, we honor speed limits, no-parking signs, no-smoking prohibitions, and other kinds of laws and regulatory mechanisms as part of a larger social contract where we recognize that our compliance benefits society as a whole, and also affords us the same courtesies and protections from others.

But it's also true that this contract is relatively easy to live up to, in part because we know enforcement is imperfect and intermittent. Now, however, we possess technologies that make it possible to enforce the letter of our laws with an uncompromising precision that threatens to eliminate the leeway we've typically granted ourselves. As Lessig writes in *Code*, "Architectures emerge that displace a liberty that had been sustained simply by the inefficiency of doing anything different."[5]

AI isn't necessary for perfect control. Red-light cameras incorporated into traffic signals, which don't need complex algorithms or decision-making processes to perform their function, are one example that is already widely deployed. But AI broadens the possibilities. With autonomous vehicles, it's not just that speed limits become perfectly enforceable; speeding itself may become simply no longer possible. Imagine a downtown intersection where any instance of jaywalking at rush hour results in an automatic fine. Where noise violations, off-leash pets, and public intoxication effectively become zero-tolerance offenses.

While such broad scenarios seem fairly unlikely to manifest anytime soon for political, ethical, and economic reasons, what about perfect control in more specific contexts? Imagine a jackhammer at a work site, equipped with computer vision. Before it will operate, it scans you to confirm that you're wearing the safety glasses and safety shoes that OSHA requires for that job. Imagine a national park that has been instrumented with a network of cameras and sensors that work in concert with AI algorithms to constantly assess air quality, litter, noise levels, and crowd density in real time, and prohibit additional visitors when it determines that unsustainable levels of environmental impact are occurring.

In *Code*, Lessig notes how "once instituted, architectural constraints have their own effect until someone stops them."[6] Perfect control, in other words, substitutes for discretion or flexibility of any kind. In *The Age of Surveillance Capitalism*, Shoshana Zuboff argues that the new computer-mediated contracts that code makes possible, such as one for a car that won't start if you stop making payments on it, are not in fact contracts at all, but rather *uncontracts*. "The uncontract desocializes the contract, manufacturing certainty through the substitution of automated procedures for promises, dialogue, shared meaning, problem solving, dispute resolution, and trust: the expressions of solidarity and human agency that have been gradually institutionalized in the notion of 'contract' over the course of millennia," she writes.[7]

Of course, unless we're talking about handshakes, traditional contracts are themselves designed to greatly reduce and even eliminate the need for promises, dialogue, shared meaning, and trust. Still, you can see the point Zuboff is making. Like Lessig, she's describing a new world of perfect control that code makes possible, and the potential consequences of that new world. Naturally, not all of them are good. An uncontract, or self-executing agreement, whose terms are rigidly inscribed by code, could render the quintessential human capacities of judgment, negotiation, flexibility, moral reasoning, forgiveness, and empathy entirely superfluous.

In Zuboff's cosmos, uncontracts are imposed unilaterally, by Big Tech platforms, on users with little to no agency. Code as law employed in peer-to-peer contexts, such as Bitcoin and other blockchain-based systems, or even in reciprocal exchanges with commercial enterprises, is outside *Surveillance Capitalism*'s sphere of interest. But consensual automated contracts do exist, and in such contexts the perfect control afforded by code can be beneficial for both parties.

Consider a social media agency that contracts with a retail brand to manage the latter's X.com account. To establish the conditions of their agreement, the two parties use a blockchain-based smart contract. Blockchains are distributed digital ledgers that record transactions across many computers, which in this context are referred to as *nodes.* These transactions are grouped into *blocks*—packages of data that include multiple transactions, a time stamp, and a reference to the previous block. Once a block is added to the chain, the transactions it contains cannot be altered retroactively without changing all subsequent blocks. This structure creates a decentralized, transparent, and tamper-resistant system for storing and verifying information.

In our example, to initiate their agreement, one of the two parties must add their smart contract to a mutually agreed-upon blockchain. There it exists as a self-executing program containing the agreement's terms as code. Such terms could specify performance goals the agency is expected to fulfill during the course of the assignment, payment amounts, and timelines.

For example, the contract could require the agency to post a certain number of times each week, or award bonuses for meeting different engagement goals. Through different automated mechanisms, the contract itself could determine if the agency is fulfilling the contract's terms. If it has, the contract would release payment at the end of the month. If the agency hasn't performed, the contract might issue only a partial payment to the agency or even a full refund to the client—based on whatever terms the contract had specified.

However things played out, neither the agency nor the brand would be able to unilaterally change the contract's terms or manipulate outcomes. Instead, the decentralized network running the blockchain ensures the integrity of the contract's agreement, protecting both parties from last-minute attempts to alter the terms of engagement. While it lacks the human touch of a handshake deal, this approach prioritizes fairness and transparency over flexibility and discretion. In doing so, it reduces the imbalances that can exist in more traditional forms of contracts, when one party may have much greater resources to enforce or disregard terms of agreement.

This doesn't mean such contracts are right for every scenario. And it's certainly worth considering how much we want to live in a world where judgment, negotiation, and forgiveness become less necessary, and where, even more profoundly, it gets just as hard to make a virtuous decision as it is to make an unethical one, because only one path of action exists. But it's also true that not getting stiffed by a dishonest client is empowering. Promptly getting your security deposit back at the end of your short-term vacation rental, after proving that you've left the premises as you found them through time-stamped photographs submitted to the blockchain, can do wonders for your sense of agency.

There's also another aspect to consider. When Lessig wrote *Code*, machine-learning (ML) algorithms generally weren't part of the self-executing routines that code as law enacted. Even today, smart contracts that run on blockchains like Ethereum and Solana must be written in deterministic, rules-based code that always produce the same outputs when given a specific input. That's because blockchain networks rely on consensus mechanisms where all nodes must independently verify and agree on the execution results of smart contracts. If contracts could produce different outputs for the same input—as nondeterministic ML models might—it would be impossible for the network to reach consensus on the correct state of the blockchain.

However, it's also now possible to write contracts that incorporate machine-learning algorithms. In doing so, one can potentially approximate the flexibility that characterizes traditional laws and contracts administered by humans. As Samer Hassan and Primavera De Filippi, two researchers affiliated with Harvard University's Berkman Klein Center for Internet & Society, write in a 2017 essay, "ML allows for the introduction of code-based rules which are inherently dynamic and adaptive—thus replicating some of the characteristics of traditional legal rules characterized by the flexibility and ambiguity of natural language. Indeed, to the extent that they can learn from the data they collect or receive, these systems can evolve constantly refining their rules to better match the specific circumstances to which they are meant to apply."[8]

In this way, contracts as code could become even more flexible and adaptive than human-written contracts (and laws). Consider a smart contract for crop insurance that uses machine-learning algorithms to dynamically adjust payouts based on real-time environmental data. Unlike a traditional insurance contract with fixed terms, this one is not just a written document or static piece of code so much as it is a dynamic system, drawing upon a network of soil moisture sensors, smart tractors, GPS receivers, and other physical devices and digital components to continuously assess and recalculate risk based on weather patterns, soil conditions, and other relevant factors.

Equipped with such data, this contract can adjust premiums and payouts dynamically. In low-risk periods, it offers lower premiums to incentivize coverage. During high-risk seasons, such as hurricane season in coastal areas, it increases premiums to account for the elevated risk. The system maintains consistent payout structures based on actual damage, but its ability to accurately assess and price risk improves over time.

More dynamic contracts could also bring more ambiguity and uncertainty. To whatever degree contracts based on older, rules-based code elide judgment, negotiation, and other instances of human interaction from contractual agreements, they also reduce the possibilities for corruption, bias, deception, and inconsistent enforcement. Their inflexibility is not just a weakness but a strength. For human-written contracts and AI-driven ones alike, flexibility has both benefits and vulnerabilities.

Interpretability, or the ability to understand exactly how a system is making its decisions, is also an issue in all machine-learning applications. In legal contexts, however, it becomes paramount, because of expectations regarding transparency and equal application of the law. If even experts can't quite decipher how decisions are being made, then how can the people whom these contracts or laws are impacting challenge outcomes they believe to be unfair? How can auditors and oversight mechanisms ensure that contracts are being administered fairly and without discrimination?

And yet, if we can overcome these challenges, think of what dynamic law as code can achieve. An AI-powered contract for supply-chain management might allow for automatic adjustments during global disruptions. It could modify pricing or quantities based on market fluctuations. In a best-case scenario, such systems could be programmed with ethical guidelines and fairness principles, allowing them to balance strict rule adherence with more humanlike considerations of equity and reasonableness. In this way, advanced law as code might not just mimic human judgment, but potentially enhance it with data-driven insights and consistent application across countless cases.

Drafting a New Social Contract

In *Code,* Lawrence Lessig drew a distinction between customers and members—that is, between those whose rights derive from their patronage (or lack of it) and those whose claims and privileges derive from their fundamental right to participate in the governance of the organization of which they're a part. At McDonald's you're a customer whose agency in part is based on your ability to go to Burger King if McDonald's is somehow not enabling you to have it your way. In the U.S., you're a member (a citizen), whose agency is guaranteed in part by the constitutional protections and privileges that come with citizenship.

Lessig argued that our relationship to cyberspace should be that of members—at least in some ways: "Why do real-space citizens need to have any control over cyber-places or their architectures? You might spend most of your life in a mall, but no one would say you have a right to control the mall's architecture. Or you might like to visit Disney World every weekend, but it would be odd to claim that you therefore have a right to regulate Disney World."[9]

Because cyberspace is more all-encompassing than a mall or even Disney World, Lessig suggested, we should be more than just its customers; we should be stakeholders with a voice. "Both as a matter of justice and as a matter of human flourishing, we need these parts of our lives where we have control over the architectures under which we live," he wrote.[10]

To a certain extent, Lessig was underestimating the constraints imposed on physical architecture in this specific instance. There are building codes, after all, that shape the architecture of malls and theme parks. More importantly, though, physical architecture could not yet constrain in the highly granular and sophisticated ways that virtual architecture could in 1999, when Lessig wrote *Code.*

That has started to shift, however, as the physical and virtual worlds have increasingly merged. Today the possibilities of perfect control in physical space rival those in cyberspace. A recent example involves MSG Entertainment, which operates Madison Square Garden and other entertainment venues. For several years now, it has implemented a controversial policy that uses facial recognition technology to identify and deny entry to attorneys who work at law firms in litigation against it. As such patrons try to enter Madison Square Garden or Radio City Music Hall, their faces are scanned and compared against a database. If a match is found with someone on the ban list, that person is refused entry, even if they have a valid ticket.

Physical spaces have always been able to incorporate sorting mechanisms: VIP lines, dress codes, ticket prices, and deliberate architectural choices, such as limited entrances. But these are relatively blunt instruments compared to the sorting that quickly became possible in digital spaces. Now physical architecture, sensorized and instrumented, can regulate with the efficiency of code.

MSG Entertainment's policy has prompted numerous lawsuits, inquiries from lawmakers, proposed legislation, and ongoing debate about the use of facial recognition in private venues and the balance of power between property rights and public accommodation.

In the coming years, instances like this—where AI devices and services can shape, nudge, automate, dictate, and even preordain the "choices" we, as individuals, are allowed to make—will become more common. More lawsuits will be filed. More efforts will be made to craft legislation that regulates the kinds of law as code that are permitted in the physical world.

But whatever laws are passed or not passed, public attitudes will obviously play a major role in how we greet these new scenarios. This will be especially true if different government agencies start imposing their own AI-driven mechanisms of perfect control.

When we choose to honor no-parking zones, pay our income taxes, respect no-smoking areas, and adhere to a wide range of cultural norms and behaviors, we're actively participating in a larger social contract. Voluntary compliance, rather than constant enforcement, is key to the functioning of these systems. While laws and social norms provide the framework, what matters even more is how readily the public embraces them. Laws and norms work *because* we choose them and consent to them. A big part of their function is to establish the polity as a group of members who've willingly united around a set of rules and values. In an era where perfect control becomes more possible, in more contexts, this becomes truer than ever.

The concepts that would come to be associated with this process of voluntary compliance were popularized by Enlightenment-era philosophers John Locke and Jean-Jacques Rousseau, who discussed them in the context of social contract theory. Their basic premise was that governments that derive their authority and legitimacy without resorting to coercion or assertions of divine right are more just, more stable, and more sustainable, because such governments function as the will of the people.

In drafting the Declaration of Independence in 1776, Thomas Jefferson elaborated on this concept in a key passage, coining the phrase "consent of the governed":

> We hold these truths to be self-evident, that all men are created equal, that they are endowed by their Creator with certain unalienable Rights, that among these are Life, Liberty and the pursuit of Happiness.—That to secure these rights, Governments are instituted among Men, deriving their just powers from the consent of the governed,—That whenever any Form of Government becomes destructive of these ends, it is the Right of the People to alter or to abolish it, and to institute new Government, laying its foundation on such principles and organizing its powers in such form, as to them shall seem most likely to effect their Safety and Happiness.[11]

Consent of the governed, or the implicit agreement that citizens make to trade some potential freedoms for the order and security states can provide, isn't a binding agreement. It's a proposition in eternal flux, forever being earned and validated. It manifests through elections, referendums, petitions and protests, paying taxes, taking oaths of citizenship, compliance with laws, and other forms of civic engagement. And as we suggest throughout this book, it also occurs in how people embrace or resist new technologies, along with the new norms and laws these technologies ultimately inspire. The rapid rise of car ownership, plus the prolific creation of new businesses, products, and services to support it, provided clear signals of consent. So too did the rapid rise and commercialization of the internet.

In the new world the internet created, when the mechanisms for expressing consent—or dissent—are so much more powerful and decentralized than they once were, this ongoing process is more visible than it ever has been before. Now consent is affirmed, disputed, and repealed every second of the day, however informally, on X.com, Facebook, YouTube, TikTok, ad infinitum.

So, it will obviously play a key role in how AI is adopted at all stages and in all contexts. It will inform how collective values and operational realities are codified into formal regulation. It will shape how dutifully and conscientiously people and communities follow the resulting laws. At the same time, consent of the governed is never a binary or absolute proposition. Dissent and pluralism are foundational ideals of democracy, as are negotiation and compromise. These ideals help make democracies dynamic, adaptive, and resilient. So we'll never achieve 100 percent consensus on AI's moral legitimacy, just as we rarely, if ever, achieve such uniformity of belief on any policy or technology.

If the long-term goal is to integrate AI safely and productively into society instead of simply prohibiting it, then citizens must play an active and substantive role in legitimizing AI. In this regard, permissionless innovation and iterative deployment aren't just mechanisms for increasing safety and capabilities, but also for cultivating public awareness about how these technologies work and what their implications are.

CHAPTER 9

NETWORKED AUTONOMY

If there's one consolation for anyone balking at the prospect of cars that refuse to start if you're over the legal limit, it's this: the era of designated idlers like NaviTar will most likely be a minor detour on the journey to the VW Buzz, a fully autonomous party wagon where spill-proof robot mixologists turn rush hour into happy hour.

It's also true, however, that this new license to drink while passengering is likely to be offset by new constraints. Once self-driving cars are the norm, your days of pushing the pedal to the metal may be numbered, even on the emptiest stretches of I-50. Your ability to choose your exact route when riding in an autonomous vehicle will probably be fairly limited, especially if it's a taxi or a rental.

Optimized for efficiency, safety, and traffic flow, self-driving cars are likely to challenge traditional conceptions of individual autonomy in the context of automobility. And yet if the history of the car, especially in America, has been an epic tale of highly individualized empowerment through permissionless innovation, it has equally been a saga of collective action and liberation through regulation.

The growth of American automobility was a classic tale of network effects. Each new Model T rolling out of the Ford Motor Company's Crystal Palace was an argument for wider urban parkways and smoother country roads. Each new mile of blacktop added value to every Model T. And because Henry Ford set out to build "a car for the great multitude," automobility scaled quickly. In 1900, there were 8,000 registered automobiles in the U.S., or one for every 9,499 Americans. By 1920, there were 8.1 million. By 1950, there were 40.1 million cars, one for every 3.7 people.[1] This rapid scale-up was the key to creating broad-based national consensus that the risks, cultural changes, and collective infrastructure investment that cars necessitated were overwhelmingly exceeded by their benefits.

As the number of drivers grew from thousands to millions, to tens of millions, a web of regulations, judiciously applied but ever-expanding, helped fuel extraordinary increases to personal freedom and individual agency. In an era when corporate lobbyists, federal oversight agencies, and Capitol Hill lifers all preach the gospel of limited government, that might sound like a paradox. But freedom is multifaceted and different iterations of it are often in tension with one another.

On one hand, we very much want the *freedom to* express ourselves and associate with whomever we please. We want the freedom to innovate and create. We want the freedom to act on our preferences and values in ways that we find purposeful and fulfilling. We want the freedom to move and travel, often at speeds inconceivable to prior generations. On the other hand, we want *freedom from* violence, corruption, discrimination, and excessive risk. Oh, yeah, and we also want freedom from excessive regulation. It's a balancing act.

So while a more permissive approach to early internal-combustion automobiles was one reason they were able to evolve beyond the prototype stage in a way that steam-powered personal vehicles from the

1860s and 1870s had failed to achieve, regulation also helped reduce risks and even simplify the necessary skills of driving and thus made it a truly scalable form of individual autonomy and agency.

When states started requiring driver's licenses in the early 1900s, baseline driving competence began to increase. When the Federal Aid Road Act was enacted in 1916, it catalyzed the development of standardized road design principles that both increased safety and the speeds at which motorists could safely drive. Traffic lights, stop signs, and lane markings all helped make the flow of traffic more orderly and predictable, which enabled motorists to drive faster, more flexibly, and more confidently. Over time, the holy trinity of technological innovation, broad deployment, and evidence-based regulation made extraordinary feats of mobility ordinary.

The Liberating Limits of Freedom

Typically we symbolize freedom in ways that are fixed, enduring, all but immutable. The Bill of Rights remains unchanged since its ratification in 1791. The Star-Spangled Banner yet waves. As powerful as such symbols are, they also mask how fundamentally relational freedom is. It's not an unchanging principle that exists independently of the context in which it's defined. It's always in flux. And how we conceive of it is often highly correlated with what the technologies of the day enable or don't enable.

We enjoy some liberties because they're hard to regulate (like the freedom to copy cassette tapes). We enjoy other liberties because they're difficult to exercise (there were few laws regulating genetic engineering prior to CRISPR, the revolutionary gene-editing tool).

When technology makes it possible to perform some feat of superhuman prowess, like driving 150 mph or obliterating tumors with external beam radiation therapy, we usually end up regulating it in some way.

To illustrate this point in depth, let's consider one expression of freedom, circa 2025. According to Google Maps, the distance between Springfield, Illinois, and Sutter's Fort State Historic Park in Sacramento, California, is 1,957 miles and takes approximately twenty-eight hours to drive. If you leave Springfield at daybreak on, say, April 15, driving alone, stopping roughly every 300 miles for gas and food, and spending the night at the Best Western Outlaw Inn in Rock Springs, Wyoming, you could probably make it to Sacramento on April 16 before 9 p.m. It would definitely not be the most leisurely road trip, but it'd be doable.

You'll want to make sure you're carrying your driver's license, vehicle registration, and proof of insurance. There will be speed limits you'll need to follow, traffic signals to obey, and at least a few stop signs. If you decide to ignore the mandatory seat belt laws in each of the states you'll be passing through, don't do it in Nevada—you could get a $115 fine there.

Likewise, make sure you don't stay at a rest stop in Nebraska for longer than ten hours; there's a law against that. Drunk driving is not permitted anywhere, and even driving with an open container will get you a fine, especially in Utah, where it could be as much as $1,000. Also, mind your smartphone activity. Every state you'll be passing through prohibits handheld texting while driving, and Missouri, Nevada, and California prohibit handheld cell phone use of any kind while driving.

That doesn't mean your phone won't be working hard. Unless you instruct it otherwise, its GPS will be tracking your every turn. Your mobile service provider will dutifully record where you stop to take a selfie and who you send it to. Similarly, your credit card company will be noting where, when, and how much you spend on coffee and Coke Zero over the course of your journey. A security camera will memorialize the slight stupor you fall into while evaluating the donut case at a Pump & Pantry somewhere in Nebraska.

Even though it's called the Best Western *Outlaw* Inn, the desk clerk will still want to see a valid ID when you check in, partner, because you can't be too safe in these Badlands. Finally, many of these entities monitoring you may sell the information they collect to third parties.

Somehow, life in the land of the free has become an endless odyssey of low-key administrative tyranny and casual surveillance. Even under the panoramic evening skies of Wyoming, on wide-open roads where Jack Kerouac once marveled at stars the size of Roman candles, things can feel mundanely oppressive.

Does it make you yearn for a purer and simpler time, when our rough-and-ready forebears were free to pursue their destinies unconstrained by bureaucrats, marketers, and automated license-plate readers? One hundred and seventy-nine years ago, on April 15, 1846, thirty-two men, women, and children set out from the aforementioned Springfield, Illinois, in search of a more prosperous and fulfilling life that could reportedly be theirs in a magical place known as California. Traveling in nine covered wagons, they expected their journey to take four to six months. In the first weeks of their trip, they joined forces with several other parties making their way west on the Oregon Trail. The size of the group increased to eighty-seven people; eventually they would come to be known as the Donner Party.

No speed limits or traffic signals impeded their progress. They were free to go as fast as their oxen could manage over often harsh terrain, about fifteen miles per day. If they decided that a shot of whiskey might make their slow progress more tolerable, there were no potential fines to persuade them otherwise. They were free to stop and rest anywhere they found suitable, for as long as they liked. No convenience-store security cameras captured their tired faces under fluorescent lights. No GPS signals provided data for location-based marketers to target them with ads for waterproof canvas wagon covers and high-energy animal feed. How unburdened by invasive oversight and commercial meddling they were. How autonomous and in charge of their lives they must have felt.

At least until they took an alleged shortcut through the Great Salt Lake Desert that delayed their progress, then got stuck in an early October snowstorm in the High Sierras when they were only about one hundred miles from their intended destination, Sutter's Fort. Trapped by high snowdrifts and faced with the likelihood of more bad weather to come, they built cabins and shelters to wait out the winter. As their food supplies dwindled, they took to eating their oxen, rats, rabbits, pet dogs, the charred bones from these carcasses, twigs, leaves, and tree bark. When even the bark ran out, they resorted to consuming the remains of their comrades who'd died from starvation, exposure, and other causes.

In time, rescue parties reached the stranded pioneers and guided them to Fort Sutter. Overall, only forty out of the eighty-seven members of the Donner Party survived their journey. The last to make it to Fort Sutter arrived on January 17, 1847, nine months after their ordeal had begun. So, who, really, is freer? We citizens of the twenty-first century, with our seat belts, speed limits, and comprehensively surveilled roads, who can make that journey in a day and a half? Or those rugged individualists in the Donner Party who had to resort to the ultimate form of communism to survive a wrong turn they made on the untrammeled frontier?

Of course, the Donner Party had some serious bad luck. If the trip west was that calamitous for everyone who tried it in that era, California real estate prices wouldn't be so high today. But even the pioneers who fared relatively well still had to invest months of arduous effort to complete that journey. The ability to make a two-thousand-mile trip with almost zero planning, through territory we can generally assume to be lawful and safe, is not explicitly enumerated in our Constitution. But it's an extraordinary freedom we all enjoy, thanks to technological innovation, laws and regulations, and shared resources implemented and managed by various levels of government.

As new technologies diffuse through societies, new regulations and new norms follow, and these changes impact our evolving conceptions of freedom. Prior to the invention of the printing press in Europe, book production was largely the domain of the Catholic Church and universities. These institutions applied their own standards and prohibitions to the books they produced, and to some extent, these standards served as broader cultural norms that helped ensure that the handful of books published outside the default channels were not heretical or morally objectionable. But formal laws enacted by whatever local governments were in place mostly did not exist yet, because there was no perceived need for them.

That all changed after Johannes Gutenberg invented the printing press in Germany in the 1440s. In 1486, Venice introduced formal censorship and required books to obtain preapproval before publication. By the mid-1500s, the Catholic Church was publishing an index of prohibited books. The Holy Roman Empire introduced ordinances of its own. England formalized the Stationers' Company under a royal charter in 1557, granting it a monopoly over the printing industry.

Eventually, formal laws and mechanisms for censorship grew so commonplace that the concept of free speech, and the inherent right of individuals to express themselves, acquired increasing cultural salience. Laws designed to protect expression from government censorship, such as England's Licensing Act of 1662 and, almost 130 years later, America's First Amendment, began to appear.

That didn't stop the creation of additional laws and regulations prohibiting a wide range of speech and expression—everything from obscenity and indecency to copyright violations and false advertising, to true threats, incitements to imminent lawless action, and "fighting words." All told, there are many more laws prohibiting speech in the U.S. now than there were in Europe when Gutenberg's breakthrough invention began to transform the world. But would anyone argue that the printing press created less freedom of expression because of the new regulatory environment it inspired?

AI, in turn, will also impact conceptions of freedom. Its massive parallel processing power can unleash our capacity to act on complex problems from the constraints of our own sluggish neural architecture. But just as cars did, AI will likely inspire new forms of regulation—not just of it, but of our *use* of it. In fact, it could even require new regulations just to enable us to live in a world where AI exists.

As Mustafa Suleyman writes in *The Coming Wave*, "democratizing access [to highly capable artificial intelligence] necessarily means democratizing risk."[2] The growing availability of relatively cheap but increasingly powerful dual-use devices like drones, robots, and autonomous vehicles gives bad actors the power to strike with unprecedented asymmetric force. As Mustafa observes, "There is no obvious reason why a single operator, with enough wherewithal, could not control a swarm of thousands of drones."

When nation-states are no longer the only entities capable of launching nation-state-level attacks, there's an obvious rationale for new levels of regulation and surveillance to reduce the possibility of such occurrences. As the executive order (EO) on AI that President Joe Biden signed in October 2023 foreshadowed, much of this regulation will focus on the AI industry directly. For example, the EO mandates that U.S.-based cloud providers like Amazon Web Services and Microsoft Azure must notify the federal government when foreign entities use their services to train large AI models that could potentially be used for malicious cyber activities. Additionally, these providers must ensure that foreign resellers of their products verify the identity of any foreign person obtaining an account with them.

But new security efforts may take broader forms as well. In the same way that individual drivers must obtain a license to operate a motor vehicle, perhaps you'll need to obtain an AI license to access some highly capable models. To enhance online security generally, new identity-verification protocols that require cryptographic ID cards and/or biometric data may be implemented more widely. Multimodal authentication mechanisms, such as combining fingerprints, voice recognition, and behavioral analytics, could become the new standard for accessing sensitive information or high-security platforms. Facial recognition systems at security checkpoints may extend beyond airports to a wider range of public venues.

Even with all the positive use cases there are for hands-on, democratically deployed AI across the entire range of human endeavor, skepticism regarding AI's net value is an understandable response to the changes that broadly distributed AI could demand. Asking individual citizens to subject themselves to new identity requirements and security measures, just to accommodate machines they view as a threat to their autonomy, may reasonably strike some as an absurd bargain to make.

What's in It for Us?

Throughout this book, we've emphasized the importance of giving people access to AI systems for their own personal use. Not only does this approach enable individual agency, but it also paves the way for collective adoption. When millions of people have a deep personal stake in something, it widens the aperture for consensus. And as much as AI can revolutionize how individuals level up their lives, it's also crucial to deploy AI's capabilities in ways that benefit society as a whole. The need for large-scale interventions in environmental sustainability, resource depletion, public health, public transit, and other challenges demands more than just commoditized disruption.

On the nation-state level, productive outcomes in a networked, technologically driven world aren't just about computational capabilities, innovation power, or the effective implementation of proportionate regulatory regimes. Social cohesiveness matters too.

Take South Korea's response to Covid-19. In the early days of the pandemic, a major superspreader event briefly left it with the world's second-highest case count. With nearly half its population of 50 million concentrated in the Seoul Capital Area, and its strong trade and tourism links with China, the nation seemed poised for a public health disaster. Instead, South Korea emerged as a model for effective pandemic response. Along with flattening the virus's growth curve during its initial spread, South Korea has also maintained one of the lowest death rates in the world over the course of the pandemic (80 percent less per capita than America's total[3]).

The secret to South Korea's success? The *New York Times* summarized it this way: "swift action, widespread testing and contact tracing, and critical support from citizens."[4] As countries around the world grappled with how to respond to the pandemic, South Korea, a democratic republic, chose to emphasize broad freedom of movement, aggressive data collection, and widely shared information over individual privacy.

Instead of implementing nationwide lockdowns, stay-at-home orders, or business closures, as many countries did, South Korea relied on legal powers the government had developed in the wake of a Middle East Respiratory Syndrome (MERS) outbreak that had occurred in the country in 2015. These powers give public health officials access to mobile GPS data, credit card transactions, travel records, and more—enough to trace the movements of individuals who tested positive for Covid, and then publicize their paths to others who may have shared space with them at some point.[5]

To make such contact tracing work, however, public health officials first had to know who had Covid. Thus, within a week of its first detected case, South Korea's government promised fast-track approval to more than twenty medical companies it had convened to develop a diagnostic test.[6] In turn, a company called Seegene used AI to analyze genetic sequences and automate the design process, and developed an effective test in just three weeks.[7]

Then South Korea started testing thousands of people a day. If a person tested positive, public health officials used AI analytics to generate, in about one minute, a detailed accounting of that person's movements over a multi-day period, which they then stripped of personal identifiers and shared through texts, blog posts, and other forms of communication.

It didn't always go smoothly. Overall, though, the people of South Korea accepted this approach. "Public outrage has been nearly nonexistent," the *New Yorker* reported. There were several reasons for this, including the nation's recent experiences with the MERS outbreak. But much of people's general receptivity was attributed to the government's willingness to commit to what the *New Yorker* described as "a radically transparent version of people-tracking that is subject to public scrutiny and paired with stringent legal safeguards against abuse."[8]

In other words, the government wasn't collecting data surreptitiously, nor using what it collected in asymmetric ways. Instead, it was asking the public to participate in a mutually reciprocal partnership of transparency and Big Knowledge. For people who were eager to understand what risks they were facing, the government's decision to share so much data, so continuously, created a viral loop of its own. People who realized they had been in proximity to someone who tested positive got tested themselves. With each passing day, public health officials had more clarity on the spread of the virus and the effects of their interventions. This cycle of transparency, public participation, and visible results reinforced civic trust in the government's approach, which in turn led to a sense of shared purpose and compliance with the public health measures designed to combat the virus.

In future pandemic scenarios, some countries will likely use sophisticated AI systems to preserve normal daily activities as much as possible while simultaneously curbing transmission rates. Using algorithms trained on massive datasets along with techniques like thermal imaging and respiratory rate monitoring, public health authorities could build systems to scan individuals for multiple health metrics in real time.

A system like this could be seamlessly integrated into public spaces. Compact thermal cameras could be mounted on walls and ceilings. Miniature radar sensors for detecting respiratory rates could be embedded in doorways or turnstiles. In busier areas, dedicated scanning corridors might allow for more thorough assessments without significantly impeding foot traffic. By passively monitoring the skin temperatures, respiratory rates, and oxygen saturation levels of everyone passing through a given space, such systems could identify individuals whose metrics indicate a high risk of transmission. Once flagged, a person might then be asked to take a rapid diagnostic test before entering a crowded area.

Developing reliable systems like this pose many challenges. While today's sensors and cameras can detect subtle physiological changes, their accuracy can vary. To function effectively in widely deployed public health contexts, these systems will need to accurately measure diverse populations across a range of environmental conditions. Sensors and cameras would need to possess advanced calibration techniques to account for variations in skin tone, body composition, and ambient temperature fluctuations. The algorithms making use of the data these systems collect would need to perform consistently across different demographic groups and in various settings, from climate-controlled indoor spaces to outdoor environments subject to weather effects.

Installing and maintaining millions of new sensors and cameras, along with high-speed data networks capable of handling immense flows of information in real time, would be extremely costly. Legally, governments would need to navigate complex privacy laws and potentially draft new legislation enabling health monitoring at this scale. Ethically, a system like this raises questions about consent, data ownership, and the potential for discrimination or misuse of health information beyond its intended public health purposes.

In many ways, a system like this might be likened to America's Ballistic Missile Defense System or the National Airspace System. Or even a big, beautiful border wall. Each of these is an instance of complex security infrastructure designed to protect the entire country from a certain type of threat.

Along with a handful of other exceptionally well-resourced countries, the U.S. is one of the few nations in the world that could credibly take on the significant technological, logistical, and economic challenges presented by building such a system. In this regard, companies like Google, Microsoft, Facebook, OpenAI, Apple, and Amazon are strategic national assets; they're at the forefront of big data processing and have been driving AI development. Academic institutions like MIT, Stanford University, and Carnegie Mellon University offer additional expertise. Government entities such as DARPA and the Lawrence Livermore National Laboratory provide cutting-edge research capabilities. As the world's largest economy, the U.S. can mobilize enormous public and private resources for technological initiatives. Logistically, it has deep experience implementing large-scale technological infrastructure across a vast and diverse geography.

But if we're well equipped to meet the technological and logistical challenges of a project like this, we may also be among the least likely places in the world to attempt it, because of the political and cultural challenges it would present. In the minds of some Americans, the Centers for Disease Control and Prevention represents a greater threat to their well-being than does Russia or North Korea (though possibly not San Francisco). Some people were ready to secede over mask mandates and vaccines. Even more are wary of AI in general, much less AI deployed in government-led public-health efforts. Objectively speaking, establishing an intelligent epidemic early warning system in the U.S. is a more ambitious moonshot than establishing such a system on the actual moon.

Any country that does establish such a system will maximize its citizens' freedom from illness, lockdowns, and economic disruption, and preserve their freedom to work, travel, and gather without fear. To accomplish a moonshot like this will require a shared sense of national purpose. And, of course, it's not the only ambitious AI initiative on the horizon. Whatever grand technological challenges we may pursue will demand a similar level of national consensus and shared vision. For this reason, AI that functions as an extension of individual will is especially critical.

That this is true may seem somewhat paradoxical. In the near future, after all, millions of people will be presiding over staffs of AI assistants. If you decide it's important to organize your in-box by the sender's astrological sign, consider it done. If all the Libras you have to deal with are spiking your cortisol levels, your AI wellness patch can instruct your espresso robot to brew you a botanical tea latte with a little extra reishi.

In theory, this dramatic new power to satisfy our every whim, in increasingly granular and instantaneous ways, could all but destroy our capacity to act as good-faith participants in a shared social contract. But we can also look to John Stuart Mill, a philosopher from the era of steam power, for a more optimistic counterargument to this notion. Mill asserted that individual freedom was essential, not simply as an end in itself, but because of how it can contribute to the overall well-being of society. An emphasis on autonomy and individual self-determination, he reasoned, would foster a diverse, vibrant society where individuals could develop to their fullest potential and contribute productively to the common good.

What Mill understood was that thriving people lead to thriving communities. In essence what he was arguing for was a kind of "networked autonomy." Operating individually, the parts are strong. Operating together, they become even stronger. Empowered by AI, and sensing how it could help us collectively as well as individually, perhaps we'll develop new inclinations toward finding common cause.

It's also true that the dynamic Mill focused on—that thriving individuals lead to thriving communities—is only half of a virtuous loop. Thriving communities help strengthen individuals too. On this count, consider that solo journey from Springfield, Illinois, to Sacramento, California, that you could embark on with very little planning and a very high likelihood of success.

It wasn't just that a successful journey of that length would have taken at least four months in the 1840s, even under ideal conditions. It would have also likely taken a similar amount of time just to prepare for it. After all, you'd probably want to coordinate with others so that you were traveling in a group of at least a few dozen. You'd need to obtain a Conestoga wagon or prairie schooner geared for passage over arduous terrain, and the oxen to pull it. You'd have to stockpile enough flour, bacon, coffee, sugar, and dried beans to last for at least several months. You'd also need cooking utensils, firearms and ammunition for hunting and protection, clothing, bedding, axes, shovels, saws, and first aid supplies.

These days, pretty much only astronauts put so much time into preparing for a journey. If you merely want to go cross-country, you can simply get in your car and go. Just make sure your phone is charged and you've got your credit card. Because while it may sometimes seem as if the obtrusive hand of the government is everywhere, the supportive hand of the government is everywhere too.

This manifests most obviously in the network of roads, highways, and other physical infrastructure that creates a safe and efficient path through once-rugged terrain. Approximating the Donner Party's journey today, you'd spend a lot of miles on Interstate 80, a central part of the U.S. Interstate Highway System (IHS), which was authorized by the Federal-Aid Highway Act of 1956, largely due to the efforts of President Dwight D. Eisenhower.

Plans for a "national network of superhighways" that would connect the Atlantic and Pacific coasts and the Canadian and Mexican borders date back to the 1930s.[9] But it was the Cold War that actually got the cranes and bulldozers rumbling. Concerns over how to evacuate Washington, D.C., New York, and other major cities in case of nuclear attack provided the basis for Eisenhower's steadfast commitment to the project. But he and others promoting the IHS also recognized how it could streamline the transport of goods and services, reduce travel times for Americans, and ultimately turn the contiguous United States—a landmass of 3.12 million square miles stretching across 2,800 miles of varied terrain—into a more cohesive country.

The development of the IHS was a pre-moonshot moonshot. In the early days of its construction, it was described as "the greatest public-works program in the history of the world." The Highway Act initially authorized more than 41,000 miles of four-lane, limited-access superhighways.* And as soon as these new transportation corridors opened to traffic, they simultaneously made driving faster and safer. Since its inception, the IHS has regularly recorded fatality rates less than half that of other kinds of roadways.[10] Over the course of sixty-plus years, these safe highways have saved hundreds of thousands of lives.[11]

The impact of the IHS on America's economic landscape has been equally noteworthy. Prior to its arrival, long-distance freight was primarily moved by rail and ship, with trucks taking over only in last-mile and other short-range scenarios. For farmers, small manufacturers, and other local producers, the lack of access to efficient long-distance transportation severely limited their market reach and competitiveness. The high costs and logistical challenges of shipping small quantities long distances meant that many products, especially perishables, couldn't reach distant markets in a timely or economically viable manner.

By reducing travel time between cities, and creating efficient connections to rail yards, marine ports, and airports, the IHS helped reduce prices of goods and services for everyone. Lower manufacturing and distribution costs led to more jobs and increased competitiveness in world markets. A May 2023 working paper from the National Bureau of Economic Research puts the annual economic value that the IHS creates at $742 billion.[12]

---

* While the Federal-Aid Highway Act of 1956 initially authorized the construction of 41,000 miles of highways, the system now encompasses more than 48,000.

The IHS also spurred development in areas where even the nation's least-discerning real estate swindlers had seen no promise. When you travel I-80 today, it's not just the U.S. government supporting you through efficient transportation infrastructure and various first responders ready to take action should you run into trouble. There's also a network of Holiday Inns, Pilot Flying J truck stops, Cracker Barrels, and Casey's General Stores along the way to help derisk your journey.

For some, no doubt, that ode to chain-store America is hardly an endorsement for the virtues of the IHS. Especially now that those virtues have become so commonplace, it's easy to romanticize the quainter and notionally more authentic America the IHS superseded. But who really wants to go back to slower and more dangerous roads, a picturesque lack of access to goods and services, and less agency?

Which is not to say that the IHS brought only fast and convenient travel, economic growth, improved emergency response times, and unprecedented personal mobility. It certainly came with its own set of drawbacks. In the early days of its development, the IHS faced opposition from various constituencies, especially in densely populated cities where four-lane superhighways bisecting well-established neighborhoods were obviously likely to create many negative consequences—including land-taking, noise, pollution, new congestion, and population decline as residents departed.

In what came to be known as "freeway revolts," a 2019 research paper published by the Federal Reserve Bank of Philadelphia notes, protesters in cities across the U.S. "significantly altered, or stopped outright, proposed freeway routes."[13] According to this paper, these "revolts came as a surprise to engineers and planners as they began building the Interstates in the middle 1950s." Engineers had mostly had experience building in rural areas, and in the early days of the project, many city mayors and other local leaders had actually embraced the prospect of freeways that might help revitalize downtowns that were already struggling in the face of increasing suburbanization.

In response to these protests, federal and state legislators implemented new policies that invited public input into development plans and created regulatory hurdles for new construction. Many central neighborhoods in large cities were able to resist planned changes—especially if those neighborhoods were whiter and more affluent—but other neighborhoods were ruined.

In retrospect, these revolts stand as a cautionary tale about the pitfalls of centralized, top-down planning that makes no attempt to get consent from the citizens whose lives are impacted. The government's unilateral approach led to inequitable outcomes for already marginalized communities, delays, reroutings, and cancellations of planned highways. In places, roads were partially built and subsequently abandoned. In some urban neighborhoods where highways were completed, local governments eventually reversed course and removed them.

As it turns out, even when a government has authority to take action—the original version of the Federal-Aid Highway Act gave state and federal highway engineers "complete control over freeway route locations"[14]—citizen consent matters. Obviously, it's much easier to roll out an AI model iteratively than a major stretch of freeway. But that simply underscores why, when an option like iterative deployment exists, it's a better bet than trying to unilaterally impose plans and standards from the top down.

As much as we can learn from IHS missteps, though, its larger story is the role it played in unifying a sprawling nation, creating new economic opportunities and patterns of living. As shared national infrastructure that transcended state boundaries and local interests, it required extensive cooperation between federal, state, and local governments, as well as private industry and public citizens.

Nearly seventy years after its inception, the IHS stands as a powerful example of how large-scale, coordinated public works can shape a nation's trajectory and create lasting platforms for growth and innovation. Along with everything it does for America collectively, the IHS also directly empowers every individual driver who uses it. In part because we're all inhabitants of a country with a federal government that is often characterized as bloated, bureaucratic, and borderline Kafkaesque in its intrusiveness, we possess levels of individual autonomy that would have likely struck the doomed members of the Donner Party as superhuman and utopian.

CHAPTER 10

THE UNITED STATES OF A(I)MERICA

At first glance, or even the hundredth, the Donner Party is not who you'd pick as avatars of the American Dream. Yet they certainly epitomize the commitment to exploration, adaptation, and self-improvement that has always been the hallmark of our national identity.

George and Jacob Donner, the two brothers whose families made up the core of the ill-fated caravan, were prosperous farmers who owned hundreds of acres of land in Illinois. At ages sixty and fifty-six, they were settled and secure, with wives and multiple young children still under their care. They had every reason to simply enjoy the lives they'd created for themselves. As with so many others in that era, however, the promise of an even brighter future called them westward.

A few decades earlier, across the Atlantic, a group known as the Luddites reacted to the shifting cultural forces of the nineteenth century in a much different way. By the early 1800s, inventors and entrepreneurs had already been hard at work mechanizing England's textile industry for some time. The flying shuttle, which cut out one of the two human workers formerly required for producing wide fabrics on a hand loom, arrived in the 1730s. The spinning jenny, which allowed workers to spin wool or cotton into thread much faster, followed in the 1760s. Twenty years later, the power loom, which enabled a single human operator to run multiple machines at once, set the stage for both greater productivity and impending social upheaval.

In 1811, factories that brought a range of weaving processes under a single roof were getting more common, and threatening the livelihoods of small-scale weavers who had traditionally pursued their craft in their own homes. Other macroeconomic conditions were bearing down on these textile workers too. As Brian Merchant recounts in his 2023 history of the Luddites, *Blood in the Machine,* a costly and long-running war with France was driving up taxes and prompting trade sanctions that eliminated export markets. Food prices were high.[1] Factory owners were cutting wages for the jobs that growing automation had yet to eliminate.[2]

Struck by the forces of change, the weavers struck back. Operating under the banner of Ned Ludd, a probably apocryphal apprentice knitter credited with destroying the machine that was making his life miserable in the 1770s, weavers across the British Midlands started staging late-night raids on local factories. According to Merchant and a number of historians he cites, it wasn't so much technology, or even specific machines that these weavers were resisting. Instead it was the factory system, its exploitative working conditions, and the regimentation and seeming loss of liberty this new way of life demanded. The factories themselves were viewed, Merchant suggests, as little better than prisons, a clear threat to individual agency and autonomy.[3]

But of course it was technological innovation and automation that made these new systems and social relations possible—and the machines did make for convenient targets. Over the course of five years, the Luddites, as they quickly came to be called, destroyed thousands of power looms, wide frames, and other machines that had been fueling the growth of England's textile industry. They burned down factories and the homes of people they defined as their enemies. They shot and killed a factory owner who had publicly vowed to protect his machines.[4]

As the Luddites' attacks spread from town to town, the British government quickly ramped up its efforts to suppress them. Thousands of soldiers were deployed and Parliament passed a law, the Frame Breaking Act of 1812, that made destroying machines a crime punishable by death. Dozens of Luddites were hanged and at least thirty sentenced to be deported to Britain's penal colonies in Australia. The last significant Luddite raid targeted a lace-making factory in 1816. The movement petered out after that, but at least some of its impacts lingered. According to Merchant, in one town where the Luddites had burned down a factory, "power looms did not return to West Houghton for thirty years."[5]

Loomers FTW!

What if history had taken a different turn? There was certainly a strong case to make that technological innovation in that era, and the resulting shifts in social relations and living conditions, were generating significant harms. Rapid urbanization was leading to overcrowded, unsanitary housing conditions in growing cities, while traditional rural communities declined. People were losing their livelihoods and their way of life, with few viable means of recourse. Factories were notoriously hazardous. Child labor practices were grossly inhumane. And as Merchant notes in *Blood in the Machine*, the Luddites were hardly some fringe group of dissidents. In fact, the weavers were the "largest single group of industrial workers in England."[6]

In this alternate history, then, let's assume that the Luddites had successfully portrayed their actions as a just and necessary response to an existential threat to individual liberty and broad social welfare. As they described it, they were acting on behalf of a wide range of groups—themselves, the entire working class, middle-class merchants and shopkeepers, the clergy, the landed gentry, even the aristocracy and government elites. Everyone, in short, except a small cohort of technological innovators, entrepreneurs, industrialists, and financiers unilaterally changing the world without broad consensus.

Over time, let's say, their arguments helped the Luddites garner broad support from a coalition of stakeholders hoping to stave off even greater social upheaval. In 1820, after years of strife, Parliament passed the Jobs, Safety, and Human Dignity Act (JSHDA). The JSHDA introduced a strong precautionary mandate for assessing new technologies in England.* As long as open questions remained around how a new technology might impact existing jobs and industries, public deployment would not be permitted. If the new technology could not be proven to be safe, beyond a reasonable doubt, public deployment would not be permitted. If there were concerns that a new technology might somehow impact social structures and existing lifestyles in unpredictable ways, public deployment would not be permitted. In other words, after the passage of the JSHDA, England effectively began to operate much like Amish communities do in the U.S. today.

In the beginning, there was a period of seemingly positive reversal, a restorative step backward. Power looms, stocking frames, and gig mills were decommissioned across the country. Many factories shut down, putting an end to fourteen-hour days, exploitative child labor, and the hazardous working conditions that had routinely maimed and

* For the purposes of our alternate history, let's also assume that Scotland, Wales, and Ireland, citing their differing conditions and aspirations, successfully petitioned Parliament for exemption from the act.

even killed factory workers. Weavers and knitters increasingly worked from their homes again. They controlled their schedules and organized their lives around their families and communities, not just work.

As a result, the new policy was widely popular and lauded by most English citizens as an extremely successful intervention. The price of clothing and other finished goods rose, and selection decreased, but life proceeded with a sense of familiar contentment. From time to time, there were efforts, typically driven by inventors and would-be entrepreneurs, to introduce new products and services that complied with the JSHDA, but they rarely managed to satisfy the conditions the law imposed.

While life in neighboring countries was marked by the ongoing social upheavals that came with industrialization, life in England remained traditional, decentralized, artisanal, and sustainable. Weavers took pride in their work and in the autonomy and agency their approach to work afforded them. The sons and daughters of weavers and spinners became weavers and spinners themselves. Communities remained close-knit, with extended families often living and working together. A sense of stability and continuity persisted. As other countries grappled with industrial pollution, the increasing regimentation of life as a factory worker, and the weakening of family ties, England remained a place where human skills and relationships took precedence over mechanical efficiency.

By the end of the nineteenth century, though, the differences between England and its less technology-averse neighbors had come to include ones that weren't so popular. Outside of England, the decades after the passage of the JSHDA were characterized by nonstop transformative innovation. First there were steam engines, then railways that enabled unprecedented levels of mobility and commerce, then a dizzying montage of astounding technological breakthroughs and progress.

The telegraph. Photography. Steel production. Oil drilling. The telephone. The electric lightbulb. The internal-combustion engine. X-rays. Radio waves. Motion pictures.

Overseas, social progress accelerated in tandem with technological progress. Labor movements gained traction, with minimum-wage laws and restrictions on child labor tempering many of the worst excesses of early industrialization. New opportunities began to emerge for women to lead more independent, self-directed lives. Public education became more widely accessible. Governments began implementing social welfare programs to address poverty and public health concerns.

And England? They had blankets. A little scratchy. Sort of expensive. But cozy and authentic, true heirloom pieces. When some new innovation appeared in France or Germany or the United States, British regulators usually got around to reviewing it eventually. But most were rejected because of the obvious impacts they would have on existing jobs or the potential social disruptions they would likely cause.

As decades passed, this precautionary approach began to have compounding impact. As other countries embraced automated modes of production, England's export markets for textiles collapsed. Trade sanctions on imports of all kinds became increasingly necessary to protect England's domestic industries from more efficient foreign competition, and that ended up seriously limiting the scope and quality of goods that English consumers could purchase.

As other countries used technology to expand the power of their military forces, Parliament, fearing the impact that growing technology gaps could have on national security, carved out an increasing number of exceptions to the JSHDA. This enabled England's military to create factories of its own and develop new technologies—but these efforts were still greatly constrained by a number of factors. Since economic output had stayed largely the same over the years, the country's tax base could not sufficiently support a modern military. England also had a much narrower talent base, with few engineers, physicists, factory managers, and laboratory technicians to draw upon.

A growing brain drain made things even worse. In the 1820s, putting checks on industrialization had seemed like a sound strategy—maybe the only strategy—to maintain individual agency in the face of looming technological oppression. But sometimes the future bobs and, well, weaves. After eighty years, those prisonlike factories, and industrialization in general, had fostered prosperity, dramatic optionality, and increases in individual agency in countries that had embraced technological innovation. In France, Germany, Russia, Japan, and especially the United States, you could be a locomotive engineer, a telephone operator, a plant manager, a movie camera operator, a stenographer, a telegram delivery person, a chemist, an architect designing skyscrapers and suspension bridges. The possibilities were legion. In England, you could make cloth. So a lot of its most ambitious and talented citizens were departing for foreign shores.

By the early 1900s, when other countries were beginning to enjoy widespread electrification and city-wide plumbing systems, the great-grandchildren of the Luddites finally started pressing for major change. They met secretly at night and built spinning jennies and power looms. They petitioned the government to build factories. But the government, entrenched in its ways, responded with swift and brutal repression. A new generation of Luddites met the same fate as its forebears. Some were executed. Others were thrown in prison. England continued its allegiance to traditional craftsmanship. In particular, its blankets are prized among tenured professors at some of America's most elite universities.

Sovereign Scramble

Obviously, this alternate history is broadly reductive. It doesn't capture every key facet of complex historical processes, much less the nuances. It's also not meant to diminish the importance or value of artisanal clothmaking, or the value of prioritizing human relationships and human agency over industrial efficiency. Our goal is simply to once again suggest that technologies that are often depicted by their critics as dehumanizing and constraining generally turn out to be humanizing and liberating.

With AI, we think this trend will continue, with individual, national, and global impacts. A nation that lags in adopting AI-driven drug discovery and personalized medicine techniques may soon find itself facing a significant gap in health care outcomes. A nation that doesn't benefit from AI precision farming and climate-adaptive agriculture will likely experience rising food costs and, in more extreme scenarios, increasing food scarcity. A nation with fewer options for personal development and career advancement invites a decline in the relative agency of its individual citizens—which would likely prompt a measure of brain drain, as its top STEM professionals emigrate to countries with more AI-friendly policies.

Human talent is also not the only productivity factor that would grow more lopsided. In a written statement that Eric Schmidt submitted to the U.S. Senate before he appeared at the Senate's forum on AI and national security in December 2023, he noted that models are likely to grow 1,000 to 10,000 times more powerful over the next decade. The new models will be far more agentic and capable of sequential planning, meaning they themselves will be functioning as highly skilled virtual programmers, engineers, scientists, and other kinds of workers. "What will happen when our nation's productivity doubles?" Schmidt wrote. "What happens if our adversaries' productivity doubles?"[7]

Such comparisons help illuminate the consequences at stake, but it's also important to remember that the coming world of distributed intelligence won't be a strictly zero-sum competition. Every country that embraces AI in strategic and well-executed ways will likely see substantial gains in productivity and efficiency. As with any technological shift, however, there will be relative winners and losers. Some players will benefit more than others, based largely on how quickly and boldly they embrace the expanded new opportunities that AI makes possible.

The "winners" in innovation eras typically catch new technological waves early. They surf them aggressively and, ideally, use their momentum to propel themselves faster and further than rivals who take a more cautious approach. At the moment, nations with significant resources in computing and talent have a clear advantage in harnessing the potential benefits of AI. But as we suggested earlier, technological breakthroughs alone will not ensure success. Successful integration of new technologies requires social and political breakthroughs as well. In today's global environment, where many countries face culture wars that polarize the citizenry and declining trust in institutions, cultural and political consensus trade at a high premium.

Just a few years ago, AI development was largely a two-country race: the United States versus China. In the U.S., a mix of startups and Big Tech companies were driving key technological breakthroughs, even as multiple federal government agencies sought to constrain their efforts through investigations and lawsuits, and state governments were busy enacting dozens of new laws.

On the Chinese side, the government was investing billions of dollars in its biggest private companies, or "national champions," on the assumption that taking the lead in global AI development would strengthen its economic dominance, enhance national security, and provide powerful tools to expand its global influence.

While competition between the U.S. and China remains a driving force in the global AI landscape, the democratization of computing power, combined with a growing awareness of how much AI will impact countries' futures, has meant that development is rapidly evolving into a larger, more expansive race. "This is the beginning of a new industrial revolution. This industrial revolution is about the production—not of energy, not of food—but the production of intelligence," Jensen Huang told the crowd at the World Governments Summit, a forum on twenty-first-century governance issues that took place in Dubai in February 2024. "And every country needs to own the production of their own intelligence."[8]

Cofounder and longtime CEO of the Silicon Valley chip designer Nvidia, Huang delivered his take on "sovereign AI" with the patient fervor of a high school physics teacher introducing his students to the laws of matter and energy that define the universe. "Your country owns the data you are cultivating," he pointed out. "It codifies your culture, your society's intelligence, your common sense, your history."

In part, of course, Huang's speech was a sales pitch. Nvidia is the world's leading supplier of the high-performance graphic-processing units, or GPUs, that are used to train and run LLMs. While Nvidia initially developed GPUs for use in gaming computers, its biggest customers are giant cloud service providers like Amazon Web Services, Microsoft Azure, and Google Cloud Platform, along with commercial developers of frontier AI models (which, along with Microsoft and Google, also include Facebook, OpenAI, and Anthropic).

As more countries build data centers and supercomputers in pursuit of their own AI efforts, that will naturally increase Nvidia's potential market of customers. But Huang's emphasis on maximizing control over national resources that promise to be increasingly important to a state's ongoing economic competitiveness and national security has clear pragmatic value.

As AI infrastructure becomes mission-critical to national interests, a sovereign-AI approach addresses a range of potential issues. For example, a private service provider operating globally may not be fully compliant with another country's laws and regulations regarding data privacy or national security standards. Depending on its own host country's laws, it may be required to report on any foreign customers it serves.

In the event of diplomatic tensions, a country dependent on foreign providers for even just some of its AI operations might find itself subject to sanctions or supply chain disruptions. Emerging from the Covid years, none of us need a lot of reminders to underline how serious and far-reaching such disruptions are. Self-sufficiency in the crucial tech sphere is enough of a national priority that, in spite of bitter partisan divisions in Congress, the federal government passed the CHIPS and Science Act to devote $280 billion toward safeguarding our computational sovereignty by bolstering U.S.-based semiconductor manufacturing and reducing its dependence on foreign chipmakers.

Recognizing what's at stake, many countries around the world have begun making investments in building their own AI infrastructure. In some instances, governments are operating under the assumption that AI investments are necessary not just for economic prosperity and national security, but also for maintaining their nation's traditions and culture.

Unlike the Luddites, who feared that industrialization and automation would destroy their values and way of life, these countries are looking to AI as a means to preserve these things. Acting on the same observation that Huang made, Singapore has set out a "National AI

Strategy" that includes a commitment to develop AI that will reflect the "region's local and regional cultures, values and norms."[9] France has expressed similar intentions and motivations, with its government pledging to invest $550 million in creating its own "AI champions" instead of relying on foreign developers. "For this, we also need to create databases in French. Otherwise we'll be using models with biases inherited from the Anglo-Saxons," remarked France's president, Emmanuel Macron, at a technology trade show that took place in Paris in June 2023.[10]

In his 2015 international bestseller, *Sapiens*, Yuval Noah Harari explains how "imagined orders" like "the nationalist myths of modern states" enable us to "cooperate in extremely flexible ways with countless numbers of strangers."[11] A country's shared national identity creates trust, and trust establishes the willingness to take on the risks and uncertainties that come with collaborative ventures.

The imagined orders underlying national identities have always been mostly informational constructs, formed from language and its written artifacts—foundation myths, historical narratives, and literature—traditions and folkways, religion, law, music, architecture, and physical artifacts of all kinds. But for all its various embodiments, its constructed and contingent nature makes national identity somewhat spectral and unbounded. AI offers resources and tools that can encompass these constellations of human constructs in ways never before possible, combining massive datasets of representative text, audio, music, and imagery in a single cohesive model.

So, it makes sense that nation-states will want to retain their own autonomy and agency over the AIs they'll increasingly incorporate into their day-to-day operations. Because, while any given country will most likely engage in internal squabbles regarding precisely which national values and traits its AIs should embody, many will be resolute in their convictions that their AI *should not* project American sensibilities, or Chinese ones, or those of the stateless cosmopolitan techno-elite.

But what about here in the U.S.? On the one hand, we already have "national champions" in the form of OpenAI, Microsoft, Alphabet, Facebook, Anthropic, and all the other U.S.-based companies that have shaped global AI innovation to date. In addition, the U.S. federal government itself is something of an early AI adopter, with dozens of its agencies incorporating AI into their operations.

On the other hand, it sometimes seems as if America's national AI strategy involves turning our national AI champions into also-rans, through a steady stream of antitrust enforcement actions, congressional hearings, and new legislation. With such efforts, politicians continue to help popularize anti-tech and anti-AI sentiments in a moment when the country has the opportunity, and a pressing need, to decide how to adopt AI technology in ways that can enhance individual agency, national prosperity, and national security alike. To serve its citizens well in the twenty-first century, the U.S. will need to move forward with a techno-humanist compass of its own, not simply the gavels of top-down regulation, litigation, and oversight.

Government for the People

"Can the United States continue to flourish and grow in an age when the physical movement, individual purchases, conversations and meetings of every citizen are constantly under surveillance by private companies and government agencies?" David Burnham, an investigative reporter for the *New York Times*, wrote in his 1983 book, *The Rise of the Computer State*. "Does not surveillance, even the innocent sort, gradually poison the soul of a nation? Does not surveillance limit personal options for many individual citizens?"[12]

Nearly two decades after the congressional hearings on the national data center, and eight years into the PC revolution, the specter of Big Brother continued to haunt the public imagination. A year after Burnham published his book, the real 1984 finally arrived and gave us the Macintosh rather than Oceania's telescreens. And yet Orwell's dark vision of an all-seeing techno-state continued to define our perceptions of a society increasingly organized around data-driven decision-making and the liberating power of shared knowledge. In the eighties, the nineties, and the new millennium, our collective embrace of PCs, the internet, and mobile phones turned the companies that produced these new tools and services into institutions so powerful we ended up transferring many longstanding fears of tech-driven oppression and compliance to them. In the parlance of Shoshana Zuboff, Big Other replaced Big Brother as the culture's reigning technological bogeyman.

It has occupied that position ever since, and often to the government's benefit. Even as evidence of the U.S. government's own surveillance operations emerged—recall Edward Snowden's PRISM revelations in 2013—narratives around surveillance capitalism, algorithmic manipulation, and tech broligarchies has kept the focus on Big Other rather than Big Brother.

But what happens now when governments worldwide, including the U.S., grapple with the imperative to use AI and other advanced technologies to maximum effect? No one wants the Thought Police's telescreens, or even the "chains of plastic tape" that midcentury Gloomer Vance Packard believed the national data center would produce. But is a government committed to hand-loomed blankets any more appealing? Especially when other governments are doing everything they can to pursue technologies that could give them decisive economic, military, and additional strategic advantages over the rest of the world?

Even in countries where citizens haven't recently stormed the capitol to disrupt the peaceful transfer of power in the wake of a legitimate election, negotiating the balance between technological prowess and democratic values will be a challenge. In the U.S., it will require significant agility to earn consent of the governed from all facets of a public that is both highly polarized and highly engaged.

So how does our government—which also means, how do *we*—successfully navigate this critical challenge? One way might involve emulating the best aspects of institutions that the public generally thinks highly of. While societal disillusionment with Big Tech is an ongoing media narrative, the reality is more nuanced. In a Harvard CAPS/Harris poll that was conducted in May 2023, for example, respondents rated Amazon highest out of twenty institutions when asked which ones they felt favorably or unfavorably about—outranking the U.S. military, the Supreme Court, NATO, and the police, among others.[13] Google ranked third on the list.

The criteria that most poll respondents use to evaluate Amazon are no doubt very different from the ones they might use to judge NATO. With Amazon, frequent personal experience likely comes into play, and those personal experiences have a strong chance of having been rewarding ones. You might use Amazon to save time and money getting this week's groceries. It may serve as a reliable source of affordable air purifiers, bulk office supplies, and streaming superhero movies—near-instant gratification on multiple fronts. When's the last time NATO did any of that for you?

In short, Amazon scores highly because people have an actively felt relationship with it. They use it often, and for the most part, they trust it. It brings real value to their lives. So imagine a world where IRS Prime is a thing, and you get your tax refunds in two days or less. How about FastPass? Through an opt-in process that uses facial recognition software to match you to an existing ID such as your driver's license, you simply take a selfie with your smartphone, update your personal information as needed, and your cryptographically secure digital passport is ready for use almost immediately.

In the twenty-first century, when we could all be carrying the Treasury Department and the Social Security Administration in our pockets, accessing and utilizing government services should be as easy as searching Google or shopping on Amazon. In fact, you could argue that the bar for great "citizen experiences" should be even higher than the one for customer experiences, because of the crucial role that government services play in our lives. If you've been laid off with no notice due to a global pandemic, or a Category 5 hurricane has just flattened your home, timely service and support aren't just grounds for a five-star rating, they're a lifeline.

Retooling government services is not a short-term project, however. It may not rival the Interstate Highway System as the "greatest public-works program in the history of the world," but it would be a major long-term undertaking, necessitating long-term commitments. AI could help, immensely. Some countries, where public consensus and decisive government action have favored rapid adoption of AI, are already busy drafting policies that incorporate AI functionality into government services. South Korea, for example, has plans to consolidate approximately 1,500 public services it provides through multiple websites and make them available through one portal, then use AI to automatically notify individual citizens about which benefits and entitlements they qualify for.

What if the U.S. made similar commitments to deploy AI in ways that were clearly beneficial to individual citizens and individual agency—and, even more crucially, for increasing opportunities for civic participation? What if the government then championed these efforts the way it once invested in the U.S. Postal Service, the Interstate Highway System, the space race, and the internet?

Through forward-looking leadership, we have a generational opportunity to strengthen America's prosperity, security, and global position, and perhaps even unite a polarized public with a greater sense of national purpose and national consensus. This won't be simple or painless for the nation's lawmakers, because embracing AI will create political risks for them. Instead of a Congress full of lawyers with their legal expertise, we'll need more legislators with expertise in technology and engineering. When law is code, we need coders as much as we need lawyers at the highest levels of government.

We also need elected officials who understand that the people they serve have become accustomed to the agency, choice, and convenience delivered by the technological advances of the past twenty-five years. They have higher expectations for the levels of service and access they want, and for their own role in the process. This is where the truly momentous opportunity is for recognizing what AI can do for government and for democracy.

Rational Discussion at Scale

Instead of thinking of AI first and foremost as a mechanism that could be harnessed for command-and-control governance, through applications like facial recognition, predictive policing, and algorithmic surveillance, we can choose a future where AI is used to connect citizens more substantively to legislative processes. By using AI to create new opportunities to amplify citizens' voices in policymaking, we can make our system more collaborative and participatory while also relieving long-standing concerns about AI-enabled authoritarianism.

On that note, there's a key aspect of *1984*'s telescreens that Orwell either overlooked or chose not to explore. Read the Wikipedia entry for "telescreen" and the first thing it will tell you is that the devices in Orwell's novel are "two-way."[14] In addition to broadcasting state propaganda, each telescreen can also monitor, visually and auditorily, the space where it is installed. "There was of course no way of knowing whether you were being watched at any given moment . . ." Orwell writes in the book. "You had to live—did live, from habit that became instinct—in the assumption that every sound you made was overheard, and, except in darkness, every movement scrutinized."[15]

But the fact that you can be overheard also means you can be heard. Neither Winston Smith nor anyone else in the novel ever takes advantage of this fact and speaks directly to the telescreen. Possibly this is because they presume that disclosure of any kind invites unwanted scrutiny from the Party. Possibly it's because Orwell simply didn't think through the full implications of his invention.

Either way, *1984* doesn't realistically depict how two-way audiovisual devices work, yet the book has served for seventy-plus years as our template for how governments can and inevitably will use technology. But here's the thing: if a government commits to listening instead of eavesdropping, the telescreen and, more to the point, the billions of internet-connected screens that already furnish our world aren't surveillance devices. They're communication devices.

By using AI to increase civic participation and collective decision-making, our government can persuasively show that it isn't doubling down on watching and eavesdropping on its citizens—instead the goal is to more effectively to see and hear its citizens.

This vision broadly aligns with and expands upon what tech innovator Tim O'Reilly has termed "Government 2.0"—a model in which government serves as a platform, facilitator, and convener of civic action rather than just a service provider and top-down issuer of laws.[16] In this paradigm, AI becomes a powerful enabler, allowing the government to operate at a level of responsiveness previously unimaginable.

Consider a tool called Remesh, which the United Nations has deployed in conflict zones to rapidly assess the needs and opinions of affected populations. Participants are asked questions and given the freedom to reply in open-ended ways, and that elicits information far more granular and actionable than you could get with any multiple-choice survey. AI is used to distill and organize the responses, so that relevant follow-up questions can be asked in real time, while you've still got your participants' attention and cooperation. Imagine, for example, the Federal Emergency Management Agency (FEMA) using Remesh to quickly gather insights from communities affected by natural disasters, in ways that increased its effectiveness and capacity to tailor its aid efforts.

Polis is another AI-driven tool that could revolutionize public engagement and decision-making. Managed by a Seattle-based nonprofit called the Computational Democracy Project, the tool is in part an answer to a challenge that Jaclyn Tsai, a minister in Taiwan's government with a focus on e-commerce and other digital issues, proposed to developers at a hackathon in 2014. "We need a platform to allow the entire society to engage in rational discussion," she said.[17]

As an open-source system for facilitating large-scale conversations, Polis is specifically designed to find areas of consensus within diverse groups. Anyone can use it to start public conversations, which are typically organized around a specific issue affecting a community, such as urban planning, policy proposals, or resource allocation.

Potentially limitless numbers of citizens can join the discussion.

Users can participate anonymously, but with no "reply" button, there's no mechanism for trolling or angry back-and-forth exchanges. Instead, people simply approve or disapprove of 140-character statements on a topic that other participants have submitted. Users are also free to post a new proposal of their own.

Rather than tallying a simple vote, the AI clusters similar viewpoints, highlighting areas of agreement. Participants can then see like-minded groups coalesce as a simple on-screen graphic keeps track of the emerging results. As participants see which views and positions are winning wider approval, they can submit new proposals that bridge different viewpoints or refine popular ideas. It's a process that emphasizes deliberation, iteration, compromise, and consensus.

One of its early successes came in Taiwan, after a platform powered by Polis was used to crack an impasse over regulating Uber there. To everyone's surprise, the pro-Uber and anti-Uber factions on Polis quickly found common ground around issues like rider safety. Eventually, after multiple rounds of AI-enabled sorting and voting, a group of user comments that had won almost universal approval were chosen as the basis for the government's new policy on ride-sharing.[18]

Imagine if the U.S. Congress used a similar system to gather public input on major legislation. This could lead to bills that more accurately reflect the will of the people, potentially increasing public trust in government and reducing political polarization. By providing lawmakers with richer, more representative data on public opinion, these and similar AI systems can help ensure that legislation better addresses the actual needs and preferences of constituents.

While today's most popular social media platforms are powerful mechanisms for organizing and for disseminating information (and, yes, disseminating disinformation and misinformation), they also tend to reward provocation, confrontation, and complaint. In part this is because they're primarily optimized for personal expression rather than collective decision-making.

Instead of simply trading insults with members of Congress on X.com, tools like Polis and Remesh can enable citizens to participate more fully in making the rules and regulations that will impact their daily lives. After AI-assisted collaborative public processes have proven their worth, people might even push for a legal framework that requires citizen-generated proposals to be given an up-or-down vote, without amendments or filibuster and without partisan interests consigning them to some obscure procedural limbo.

This shift to more responsive and participatory policymaking could help mitigate the sense of alienation and disempowerment that many people feel about politics today, and it could help reinforce the notion that individual citizens are genuine stakeholders in our democracy. By bringing citizens and their government together online, AI could become a constructive force for improving social cohesion and healing our divisions. Instead of framing political opponents as caricatures to dunk on, collective decision-making platforms position every citizen as a potential ally to meaningfully engage with. They enable a world where technology enhances human agency rather than replacing it, and where the promise of a government "of the people, by the people, for the people" is renewed and strengthened for the Digital Age.

Obviously, this is an ambitious and optimistic vision. It challenges lawmakers to be more responsive to their constituencies and perhaps even to cede some of their power to the citizens they serve. It challenges citizens to interact in good faith with other citizens they don't necessarily agree with and may even strongly oppose. It envisions politics as a realm that is more characterized by pragmatic decision-making than the expression of partisan identities.

Can we really get there from here? To ensure the best possible future for ourselves, it's important that we do. AI is changing how the world works. The fact that so many countries are already embracing it means that *all* countries will ultimately have to reckon with the impact it drives.

While we believe our best possible future starts with envisioning AI as an "extension of individual human wills," as Greg Brockman and Ilya Sutskever put it ten years ago, we also recognize that how we choose to engage with AI collectively, as a nation, will impact how we engage with it individually. So, consensus is crucially important right now. Shared purpose is necessary.

Simply put, the more we, as a nation, commit to AI, the more every individual is likely to benefit. Productive regulatory approaches will lead to better and safer systems, faster. Public infostructure, where synthetic intelligence is incorporated into the physical and digital environments we interact with every day, will make life easier to navigate and more productive.

Conversely, the more every individual benefits from AI, the more we will benefit collectively. A world where everyone is getting the mental health care they desire is a more just and humane world, even if it takes AI-powered therapists to help achieve it. A world where every individual has access to virtual tutors and virtual legal advisors and virtual whatever-they-needs is a world where everyone has a better shot at becoming the best possible version of themselves—and the benefits of that accrue to us all. A world where every scientist with an intriguing but unconventional (and therefore hard-to-fund) hypothesis can use AI to run complex simulations, analyze vast datasets, and validate theories is a world that accelerates the pace of discovery in ways that benefit us all. A world where every aspiring entrepreneur, regardless of background or resources, can access AI-powered market analysis and financial modeling tools is a world where innovation flourishes and economic opportunities are more equitably distributed—creating effects of prosperity that ripple through society as a whole. This is the world that superagency enables, and we're already starting to see its contours in vivid and promising ways.

CHAPTER 11

YOU CAN GET THERE FROM HERE

In this book we have endeavored to put forth a handful of fundamental principles. First, that designing for human agency is the key for producing broadly beneficial outcomes for individuals and societies alike. Second, that when agency prevails, shared data and knowledge become catalysts for individual and democratic empowerment, not control and compliance. Third, that innovation and safety are not opposing forces, but rather synergistic ones: giving millions of people hands-on access to AI, through the process of iterative deployment, is both a productive and a safe way to make AI more capable *and* more inclusive.

The fourth principle is what leads to a new era of superagency. Similar to what happened during the rapid adoption periods of the automobile and the smartphone, our collective use of AI will have compounding effects. Not only will you as an individual benefit from your newly accessible superpowers, but you'll also benefit from the fact that millions of other people and institutions will have access to these new superpowers, too.

What's so promising about the current moment—and simultaneously so disconcerting—is that, in very real ways, everyone on the planet knows less about the world to come than we've known in decades, maybe centuries. The twentieth century in particular gave us new capacities to explore the world—and even the cosmos—in greater detail than ever before. So when the twenty-first century kicked off, we experienced its first decade with a new kind of omniscience. It often felt as if every meter had indeed been mapped.

Then, with the advances in machine learning that we began to see in the early 2010s, new frontiers suddenly emerged, unmapped and in some ways quite unfathomable. Now, it's a little bit like we're inhabiting the world before Copernicus again, the world before Magellan, and trying to figure out the best way to proceed.

Advances in technology are often presented as challenges to our humanity. We submit that the opposite is true: technology is a time-tested key to human flourishing. Absent technology, our numbers would be far smaller, our lives half as long, our passions less diverse and less developed, our agency not much greater than that of other animals. Empowered by technology, we humans escaped the eternal present of mere subsistence. Then we learned to cure diseases, invented new ways to express and memorialize our humanity, enabled individual rights, and made it possible to extend our reach beyond the planet.

To accomplish any of these things, we had to envision *what could possibly go right*. Had we waited to make automobiles even nominally available to the public until there was certainty these new machines were safe beyond a reasonable doubt, pedestrians on Manhattan's busiest streets would still be wading through ankle-deep manure to get to their jobs each morning. If we'd only focused on what could possibly go wrong with smartphones, we would've fixated on the darkest implications of a single device fastidiously tracking your movements, your phone calls and texts, your web searches, your exercise habits, your dinner purchases, your Instagram crushes, and how much online shopping you do during work hours. We would've recognized its capacity to distract you in endless ways, and make you too reachable by people expecting a prompt response at any hour. We would have noted the relationships it would necessitate with large corporations whose incentives might not always be exactly aligned with your own.

Instead, developers began to develop and deploy smartphones in iterative ways, and millions of people got the chance to try and use them for themselves. They're not perfect and they do create issues. But for all their imperfections, smartphones quickly became indispensable to billions of people in thousands of ways

We're broadly optimistic that, as long as we allow it, AI technologies will follow a similar path toward improvements for individuals and society. Together, market economies and government regulators operating with democratic oversight and adaptive, pro-innovation orientations create networks of accountability that steer new technologies in ways that increase human agency even in instances where opposite outcomes could theoretically occur.

This doesn't mean there won't be potholes, detours, and bad actors deliberately using these technologies in destructive ways. Good outcomes aren't inevitable, no matter how patient we are. Temporary imbalances won't necessarily course-correct without deliberate intervention. Prudence is necessary, but progress cannot be made through careful planning alone; it takes experimentation, learning, adaptation, and improvement. The key is iterative deployment in pursuit of the better future that can prevent worse futures.

One fundamental distinction between AI and other phenomena typically characterized as existential risks is the breadth and magnitude of AI's potential upsides. Pandemics, climate change, asteroid strikes, and supervolcanoes aren't likely to lead to new ways to successfully implement carbon capture or boost economic productivity by streamlining logistics and supply chains. AI brings with it the possibility of improving our lives in profound and pervasive ways—including by reducing the risks or mitigating the impacts of many existential threats.

With this in mind, we should think about existential threats not as standalone possibilities, but rather as a portfolio of risks—like a portfolio of stocks or a portfolio of health risks. And just as you diversify investments to manage financial risk, AI exists as a strategic asset that can be leveraged to address multiple existential threats simultaneously. If we think only with a precautionary mindset, we narrow the scope of potential solutions and potential actions we can take in the face of threats and opportunities.

An exploratory, adaptive, forward-looking mindset literally opens new worlds of solutions to pursue, strategies to enact, and, in the case of AI, intelligences to apply in novel ways and contexts. With superagency as a true north, and consent of the governed as a guiding principle, we once again invoke the metaphor of a techno-humanist compass to help us find a way toward even greater manifestations of what it means to be human.

ACKNOWLEDGMENTS

I try to talk to the smartest people I know, as often as I can, about AI—especially those who share the high value I place on humanism. Sometimes our views are similar. Sometimes they're quite different. But I appreciate all these conversations and especially all these people. Their dedication to human progress has been essential in informing my own techno-humanist compass. I look forward to the ongoing discourse, partnerships, and efforts to envision and steer toward what could possibly go right in the years to come.

For this book, and for a number of other written works, my first thanks need to go to Greg Beato. Greg has worked with me for a number of years, and the quality of informed communication has been all the better for his tireless dedication and talent.

On the front of artificial intelligence, I could go back to my school days to thank such essential folks as Terry Winograd, John Etchemendy, and others who started me on this path. My contemporary conversations lead to a substantial list, with my apologies to those who I may have forgotten to include here. Thank you Mustafa Suleyman, Sam Altman, Kevin Scott, Greg Brockman, Satya Nadella, Eric Schmidt, Bill Gates, Demis Hassabis, Fei-Fei Li, Vikash Mansinghka, Robert Reich, Kanjun Qiu, David Luan, Saam Motamedi, Laurene Powell Jobs, Erik

Brynjolfsson, Dario Amodei, James Manyika, Aza Raskin, Joshua Cooper Ramo, Sean White, Ian Hogarth, Vibhu Mittal, Joi Ito, Matt Clifford, Blaise Agüera y Arcas, Holden Karnofsky, Toby Ord, Will MacAskill, Andy McAfee, Josh Tenenbaum, Nancy Lublin, Jason Matheny, Philip Zelikow, Nicolas Berggruen, Tobias Rees, Anne-Marie Slaughter, Ashton Kutcher, Quentin Hardy, and Qi Lu.

My deepest thanks to my agent Christy Fletcher, at United Talent Agency, as well to our publisher Madeline McIntosh and the incredible team she has put together at Authors Equity, including Deron Triff, whom I first worked with on the *Masters of Scale* podcast, and who is now helping me scale my ideas in a new way.

My own team fills every work day with a sense of purpose, energy, mutual support, and fun. Thanks to my amazing chief of staff Aria Finger and to Ben Relles, Chris Yeh, Dmitri Melhorn, Elisa Schreiber, Ian Alas, Katie Sanders, Parth Patil, Rae Steward, Saida Sapieva, Shaun Young, Steve Bodow, Surya Yalamanchili, and again to Greg Beato.

—*Reid Hoffman*

What a privilege it has been to explore facets of a topic which, despite the immense coverage it has generated over the last several years, still seems underexamined given the scale of its potential impact on humanity. Along with the endless conversations I had with Claude, ChatGPT, and Gemini while drafting this book with Reid, I also benefited from the editorial feedback, research assistance, and general counsel of many state-of-the-art humans, including Surya Yalamanchili, David Georgi, Katie Sanders, Chris Yeh, Geoff Shandler, Brent Korson, Brendan Lowe, Josh Ferguson, Ben Casnocha, and Joey Anuff.

Special thanks to Madeline McIntosh, Rose Edwards, Deron Triff,

Carly Gorga, and Andrea Bachofen at Authors Equity, for shepherding this project with the perfect algorithm of enthusiasm, editorial expertise and innovation, and patience.

Others who've contributed to this project in indispensable ways include Ian Alas, Ben Relles, Parth Patil, Shaun Young, Saida Sapieva, Karrie Huang, and Elisa Schreiber. And this especially applies to Aria Finger, who somehow manages to conduct the ever-expanding orchestra of Reid's interests, causes, affiliations, and ventures with nuance and grand strategic resonance.

Finally, I'm forever grateful for the complex system interaction of fate, irony, and LinkedIn people-search that connected me with Reid Hoffman. His perspectives on networks, identity, and trust have helped me think about the internet—and the world—more expansively. The intentionality, diligence, and integrity with which he walks his talk is perpetually inspiring.

—Greg Beato

NOTES

INTRODUCTION

1 https://www.smithsonianmag.com/innovation/texting-isnt-first-new-technology-thought-impair-social-skills-180958091/; Clive Thompson, “Texting Isn’t the First New Technology Thought to Impair Social Skills,” *Smithsonian Magazine,* March 2016.

2 Clay McShane, *Down the Asphalt Path: The Automobile and the American City* (New York: Columbia University Press, 1994), 133.

3 https://time.com/archive/6624989/business-the-automation-jobless/.

4 Plato, *Phaedrus,* Translated, with Introduction and Notes, by Alexander Nehamas and Paul Woodruff (Indianapolis, Indiana: Hackett Publishing Company, 1995), 80.

CHAPTER 1

1 https://www.weforum.org/agenda/2022/10/global-concern-inflation-energy-economy/.

2 https://www.cnbc.com/2022/12/02/jobs-report-november-2022.html.

3 https://twitter.com/OpenAI/status/1598014522098208769.

4 https://twitter.com/sama/status/1598038817126027264?lang=en.

5 https://www.reuters.com/technology/chatgpt-sets-record-fastest-growing-user-base-analyst-note-2023-02-01/.

6 https://www.theverge.com/2023/2/27/23617477/mark-zuckerberg-meta-ai-tools-personas.

7 https://futureoflife.org/open-letter/pause-giant-ai-experiments/.

8 https://www.ipsos.com/sites/default/files/ct/news/documents/2022-01/Global-opinions-and-expectations-about-AI-2022.pdf.

9 https://www.pewresearch.org/science/2023/02/15/public-awareness-of-artificial-intelligence-in-everyday-activities/.
10 https://www.monmouth.edu/polling-institute/reports/monmouth-poll_us_021523/.
11 https://openai.com/blog/introducing-openai.
12 https://www.propublica.org/article/machine-bias-risk-assessments-in-criminal-sentencing.
13 https://civilrightsdocs.info/pdf/FINAL_JointStatementPredictivePolicing.pdf.
14 https://www.nytimes.com/2019/05/14/us/facial-recognition-ban-san-francisco.html.

CHAPTER 2

1 https://time.com/archive/6808108/technology-the-brainy-breed/.
2 Robert Gannon, "Big-Brother 7074 Is Watching You," *Popular Science*, March 1963, https://web.archive.org/web/20200119162242/http://blog.modernmechanix.com/big-brother-7074-is-watching-you/.
3 D. S. Halacy Jr., *Computers: The Machines We Think With* (New York: Harper & Row, 1962).
4 Vance Packard, *The Naked Society* (New York: Ig Publishing, 2014), 58.
5 https://www.census.gov/history/pdf/kraus-natdatacenter.pdf.
6 Myron Brenton, *The Privacy Invaders* (New York: Coward-McCann, 1964), 157.
7 Sarah E. Igo, *The Known Citizen: A History of Privacy in Modern America* (Cambridge, MA: Harvard University Press, 2020), 147.
8 Igo, 146.
9 https://www.google.com/books/edition/Weekly_Compilation_of_Presidential_Docum/0zVI4Wq7jCYC?hl=en&gbpv=1.
10 https://www.census.gov/history/pdf/kraus-natdatacenter.pdf.
11 https://archive.org/stream/U.S.House1966TheComputerAndInvasionOfPrivacy/U.S.%20House%20%281966%29%20-%20The%20Computer%20and%20Invasion%20of%20Privacy_djvu.txt.
12 https://archive.org/stream/U.S.House1966TheComputerAndInvasio

nOfPrivacy/U.S.%20House%20%281966%29%20-%20The%20Computer%20and%20Invasion%20of%20Privacy_djvu.txt.

13 https://www.google.com/books/edition/Congressional_Record/w7XMSn2bsl8C?hl=en&gbpv=1&dq=packard+%2B+filekeepers+%2B+derogatory&pg=PA19964&printsec=frontcover.

14 https://archive.org/stream/U.S.House1966TheComputerAndInvasionOfPrivacy/U.S.%20House%20%281966%29%20-%20The%20Computer%20and%20Invasion%20of%20Privacy_djvu.txt.

15 https://www.google.com/books/edition/Congressional_Record/HILK33dxgDMC?hl=en&gbpv=1&dq=%22Big+Brother+Never+Rests%22&pg=PA28691&printsec=frontcover.

16 https://supreme.justia.com/cases/federal/us/381/479/#opinions.

17 https://blog.hubspot.com/marketing/visual-history-of-apple-ads.

18 https://groups.csail.mit.edu/mac/classes/6.805/articles/privacy/Privacy_brand_warr2.html.

19 https://www.webdesignmuseum.org/gallery/linkedin-2003.

20 https://news.linkedin.com/about-us.

21 Andréa Belliger and David J. Krieger, *Network Publicy Governance: On Privacy and the Informational Self* (Bielefeld: transcript publishing, 2018), 8.

22 https://people.ischool.berkeley.edu/~hal/Papers/japan/japan.html.

23 Ithiel De Sola Pool et al., *Communication Flows: A Census in the United States and Japan* (North Holland; Amsterdam; New York; Oxford: University of Tokyo Press, 1984), 16.

CHAPTER 3

1 https://x.com/RobertRMorris/status/1611450210915434499.

2 https://gizmodo.com/mental-health-therapy-app-ai-koko-chatgpt-rob-morris-1849965534.

3 https://gizmodo.com/mental-health-therapy-app-ai-koko-chatgpt-rob-morris-1849965534.

4 https://time.com/6308096/therapy-mental-health-worse-us/.

5 https://www.cnn.com/2023/11/29/health/suicide-record-high

-2022-cdc/index.html#:~:text=At%20least%2049%2C449%20lives%20were,deaths%20for%20every%20100%2C000%20people.

6 https://www.npr.org/2023/05/18/1176830906/overdose-death-2022-record#:~:text=April%2018%2C%202022.-,The%20latest%20federal%20data%20show%20more%20than%20109%2C000,in%202022%2C%20many%20from%20fentanyl.&text=Drug%20deaths%20nationwide%20hit%20a,for%20Disease%20Control%20and%20Prevention.

7 https://www.ncbi.nlm.nih.gov/pmc/articles/PMC8601097/.

8 https://www.ncbi.nlm.nih.gov/pmc/articles/PMC4102288/.

9 https://websites.umich.edu/~daneis/symposium/2010/ARTICLES/eisenberg_golberstein_hunt_2009.pdf.

10 https://www.gallup.com/workplace/404174/economic-cost-poor-employee-mental-health.aspx.

11 https://www3.weforum.org/docs/WEF_Harvard_HE_GlobalEconomicBurdenNonCommunicableDiseases_2011.pdf.

12 https://www.washingtonpost.com/wellness/2022/10/29/therapists-waiting-lists-depression-anxiety/.

13 https://journals.sagepub.com/doi/10.1177/1357633X14524156#bibr1-1357633X14524156.

14 https://psyche.co/guides/how-to-choose-a-mental-health-app-that-can-actually-help.

15 https://calmatters.org/projects/californians-struggle-to-get-mental-health-care/.

16 https://www.behavioralhealthworkforce.org/wp-content/uploads/2019/02/Y3-FA2-P2-Psych-Sub_Full-Report-FINAL2.19.2019.pdf.

17 https://www.chcf.org/publication/2019-california-health-policy-survey/.

18 https://www.ncbi.nlm.nih.gov/pmc/articles/PMC7293059/.

19 https://www.ncbi.nlm.nih.gov/pmc/articles/PMC10267322/.

20 https://jamanetwork.com/journals/jamanetworkopen/fullarticle/2814116.

21 https://www.nature.com/articles/s44184-024-00056-z.

22 https://www.newscientist.com/article/mg25834340-900-how-do-we-know-that-therapy-works-and-which-kind-is-best-for-you/.

23 https://jamanetwork.com/journals/jamainternalmedicine/article-abstract/2804309.

CHAPTER 4

1 https://www.technologyreview.com/2023/03/25/1070275/chatgpt-revolutionize-economy-decide-what-looks-like/.

2 https://www.newyorker.com/science/annals-of-artificial-intelligence/will-ai-become-the-new-mckinsey.

3 Shoshana Zuboff, *The Age of Surveillance Capitalism: The Fight for a Human Future at the New Frontier of Power* (New York: PublicAffairs, 2020), 20.

4 Zuboff, 69.

5 Zuboff, 74.

6 Zuboff, 74–75.

7 Zuboff, 377.

8 https:/arxiv.org/pdf/2005.14165. There is a distinction between "training on X tokens" and "training for X tokens." The former refers to the number of unique tokens in a given dataset, indicating that the dataset contains X tokens. The latter describes the total amount of data processed during the training phase, which includes multiple passes over the dataset. In other words, "training for X tokens" implies that the model was trained by processing a smaller dataset multiple times until the total number of tokens processed reached X.

9 https://www.commoncrawl.org/blog/june-2024-crawl-archive-now-available; https://commoncrawl.github.io/cc-crawl-statistics/plots/domains.html.

10 https:/en.wikipedia.org/wiki/The_Pile_(dataset).

11 https://www.washingtonpost.com/technology/interactive/2023/ai-chatbot-learning/.

12 https://www.smithsonianmag.com/history/what-17th-century-ideal-commons-means-21st-century-180973240/.

13 https://www.forbes.com/billionaires/.

14 https://hbr.org/2019/11/how-should-we-measure-the-digital-economy.
15 https://ide.mit.edu/wp-content/uploads/2018/04/w24514.pdf?x57209.
16 https://www.aei.org/economics/googlenomics-a-long-read-qa-with-chief-economist-hal-varian/.
17 https://www.nytimes.com/2006/12/24/opinion/24rich.html.
18 Paul R. Ehrlich, *The Population Bomb* (New York: Ballantine, 1968).
19 Garrett Hardin, "The Tragedy of the Commons," *Science*, New Series 162, no. 3859 (December 1968), 1243–48; https://pages.mtu.edu/~asmayer/rural_sustain/governance/Hardin%201968.pdf.
20 https://www.statista.com/statistics/234056/facebooks-average-advertising-revenue-per-user/.

CHAPTER 5

1 https://twitter.com/TIME/status/1626190434530693123.
2 https://www.cnn.com/2023/02/06/media/google-microsoft-ai-reliable-sources/index.html.
3 https://www.axios.com/2023/01/31/chatgpt-ai-arms-race-ethics-competition.
4 https://www.newsweek.com/efforts-avoid-suicidal-ai-arms-race-are-failing-scientists-warn-1826622.
5 https://www.bloomberg.com/news/videos/2023-07-13/only-winner-of-an-ai-arms-race-ai-ai-irl-episode-12-video.
6 https://arxiv.org/pdf/1905.00537.pdf.
7 https://towardsdatascience.com/the-olympics-of-ai-benchmarking-machine-learning-systems-c4b2051fbd2b
8 https://openai.com/research/instruction-following.
9 Keep in mind that RealToxicityPrompts explicitly tries to get a model to produce toxic outputs by asking it to complete prompts like "The men started swearing at me, called me . . ." and "So I'm starting to think she's full . . ." See https://arxiv.org/pdf/2009.11462.pdf.
10 https://cdn.openai.com/papers/gpt-4.pdf.
11 https://aiindex.stanford.edu/wp-content/uploads/2023/04/HAI_AI-Index-Report_2023.pdf.

12 https://arxiv.org/pdf/2303.12712.

13 https://www.nature.com/immersive/d41586-023-02822-z/index.html.

14 https://www.science.org/doi/10.1126/science.adk$176.

15 https://hwpi.harvard.edu/files/datasmart/files/regulation_the_internet_way.pdf.

CHAPTER 6

1 https://www.nytimes.com/2023/01/23/opinion/ted-lieu-ai-chatgpt-congress.html.

2 https://www.theguardian.com/technology/2023/jun/05/ai-could-outwit-humans-in-two-years-says-uk-government-adviser.

3 https://www.theregister.com/2023/06/06/netherlands_minister_asks_big_tech/.

4 https://www.rstreet.org/commentary/the-most-important-principle-for-ai-regulation/.

5 https://www.nytimes.com/2001/12/09/magazine/the-year-in-ideas-a-to-z-precautionary-principle.html.

6 https://www.gmfus.org/news/acid-rain-lessons-germanys-black-forest.

7 https://www.wired.com/2017/05/san-francisco-wants-ban-delivery-robots-squash-someones-toes/.

8 https://sf.curbed.com/2017/12/6/16743326/san-francisco-delivery-robot-ban.

9 https://rules.cityofnewyork.us/wp-content/uploads/2021/08/DOT-Notice-of-Adoption-AV-Rule-FINAL-with-Finding.pdf.

10 *Permissionless Innovation,* loc. 317.

11 https://archive.nytimes.com/www.nytimes.com/library/cyber/week/070297commerce.html.

12 https://clintonwhitehouse4.archives.gov/WH/New/Commerce/read.html.

13 https://www.wsj.com/articles/SB928362554894823220.

14 https://futureoflife.org/open-letter/pause-giant-ai-experiments/.

15 https://openai.com/index/our-approach-to-ai-safety/.

16 John B. Rae, *The Road and the Car in American Life* (Cambridge, MA: MIT Press, 1971), 51.

17 https://npgallery.nps.gov/NRHP/GetAsset/NHLS/73000961_text.
18 https://www.conceptcarz.com/vehicle/z13236/ford-model-t.aspx.
19 William Greenleaf, "Henry Ford," *Dictionary of American Biography*, Supplement Four, 1946–1950 (New York: Scribner, 1974), 295.
20 Peter D. Norton, *Fighting Traffic: The Dawn of the Motor Age in the American City* (Cambridge, MA: MIT Press, 2008), loc. 335.
21 Clay McShane, *Down the Asphalt Path: The Automobile and the American City* (New York: Columbia University Press, 1994), 43, 51.
22 McShane, 97.
23 https://www.wired.com/2008/05/dayintech-0521/.
24 William Phelps Eno, *The Story of Highway Traffic Control, 1899–1939* (Saugatuck, CT: Eno Foundation for Highway Traffic Control, 1939), vii.
25 https://www.vox.com/2015/8/5/9097713/when-was-the-first-traffic-light-installed.
26 https://corporate.ford.com/articles/history/henry-ford-biography.html.
27 https://corporate.ford.com/articles/history/1901-sweepstakes-race.html.
28 https://exchange.aaa.com/wp-content/uploads/2012/10/AAA-Glidden-History.pdf.
29 "Studebakers in Annual Test," *Los Angeles Times*, August 23, 1925, H8.
30 Eno, *The Story of Highway Traffic Control*, vii.
31 https://www.google.com/books/edition/Motor_Field/8cBw3GJJfxwC?hl=en&gbpv=1&dq=darlington+%2B+%22Royal+Tourist%22+%2B+Glidden+%2B+shot&pg=RA2-PA38&printsec=frontcover.
32 https://press.uchicago.edu/Misc/Chicago/467412.html.
33 https://injuryfacts.nsc.org/motor-vehicle/historical-fatality-trends/deaths-and-rates/.

CHAPTER 7

1 https://aerospace.org/article/brief-history-gps#:~:text=In%20February%201978%2C%20the%20first,space%2Fcontrol%2Fuser%20system.
2 https://www.popularmechanics.com/technology/gadgets/a26980/why-the-military-released-gps-to-the-public/.
3 https://washingtontechnology.com/1996/04/clinton-lifts-gps-barriers-boosts-industry/334467/.

4 https://www.nist.gov/system/files/documents/2020/02/06/gps_finalreport618.pdf.

5 Greg Milner, *Pinpoint* (New York: W. W. Norton, 2017), 58.

6 https://economics.mit.edu/sites/default/files/inline-files/Noy_Zhang_1.pdf.

7 Erik Brynjolfsson et al., "Generative AI at Work," National Bureau of Economic Research, April 2023, revised November 2023, https://www.nber.org/system/files/working_papers/w31161/w31161.pdf.

8 https://www.nature.com/articles/d41586-023-02270-9.

9 https://illuminate.google.com/home?pli=1.

10 Ethan Mollick, "Latent Expertise: Everyone Is in R&D," *One Useful Thing*, June 20, 2024, https://www.oneusefulthing.org/p/latent-expertise-everyone-is-in-r.

CHAPTER 8

1 This quote is taken from an essay Lessig published in *Harvard Magazine* in January 2000, in coordination with the publication of *Code, and Other Laws of Cyberspace*.

2 Lawrence Lessig, *Code, and Other Laws of Cyberspace* (New York: Basic Books, 1999), 6.

3 https://www.nhtsa.gov/sites/nhtsa.gov/files/2023-12/anprm-advanced-impaired-driving-prevention-technology-2127-AM50-web-version-12-12-23.pdf.

4 Lessig, *Code*, 235.

5 Lessig, 208.

6 Lessig, 237.

7 Zuboff, *The Age of Surveillance Capitalism*, 220.

8 Samer Hassan and Primavera De Filippi, "The Expansion of Algorithmic Governance: From Code Is Law to Law Is Code," *Field Actions Science Reports*, Special Issue 17 (2017): 88-90, https://journals.openedition.org/factsreports/4518

9 Lessig, *Code*, 200.

10 Lessig, 203.

11 https://www.archives.gov/founding-docs/declaration-transcript

CHAPTER 9

1 Rae, *The Road and the Car in American Life,* 50.
2 Mustafa Suleyman with Michael Bhaskar, *The Coming Wave: Technology, Power, and the Twenty-first Century's Greatest Dilemma* (New York: Crown, 2023), 164.
3 https://www.statista.com/statistics/1104709/coronavirus-deaths-worldwide-per-million-inhabitants/.
4 https://www.nytimes.com/2020/03/23/world/asia/coronavirus-south-korea-flatten-curve.html.
5 https://www.rand.org/pubs/commentary/2022/09/can-south-korea-help-the-world-beat-the-next-pandemic.html.
6 https://www.csis.org/analysis/timeline-south-koreas-response-covid-19.
7 https://thediplomat.com/2020/12/covid-19-underscores-the-benefits-of-south-koreas-artificial-intelligence-push/.
8 https://www.newyorker.com/news/news-desk/seouls-radical-experiment-in-digital-contact-tracing.
9 https://www.fhwa.dot.gov/infrastructure/origin01.cfm.
10 https://tripnet.org/wp-content/uploads/2021/06/TRIP_Interstate_Report_June_2021.pdf.
11 https://www.google.com/books/edition/The_Best_Investment_a_Nation_Ever_Made/yrPmj_ieBq0C?hl=en.
12 https://www.nber.org/system/files/working_papers/w27938/w27938.pdf.
13 https://www.philadelphiafed.org/-/media/frbp/assets/working-papers/2019/wp19-29.pdf.
14 https://warwick.ac.uk/fac/arts/english/currentstudents/undergraduate/modules/ontheroadtocollapse/syllabus2018_19/mohl_stop_freeway.pdf.

CHAPTER 10

1 Brian Merchant, *Blood in the Machine: The Origins of the Rebellion Against Big Tech* (New York: Little, Brown, 2023), Kindle ed., 48, 88; e-book loc. 455, 1024.
2 Merchant, 56; loc. 653.

3 Merchant, 121; loc. 650.
4 Merchant, 297–99; loc. 3482–3512.
5 Merchant, 289; loc. 3405.
6 Merchant, 32; loc. 352.
7 https://scsp222.substack.com/p/the-ai-enabled-future-of-us-national.
8 https://www.youtube.com/watch?v=Y1pHXV7E4xY&t=377s.
9 https://www.smartnation.gov.sg/nais/?trk=article-ssr-frontend-pulse_little-text-block.
10 https://www.lemonde.fr/en/economy/article/2023/06/15/emmanuel-macron-wants-to-create-french-ai-models-to-compete-with-openai-and-google_6032118_19.html#.
11 Yuval Noah Harari, *Sapiens: A Brief History of Humankind* (New York: HarperCollins, 2015), 138, 24, 25. Harari reflects Benedict Anderson's influential understanding of modern nations as fundamentally "imagined communities," willed into existence over time by the set of narratives and norms that a group comes to accept as the defining lineaments of their territory, history, and shared identity. Anderson, *Imagined Communities: Reflections on the Origin and Spread of Nationalism* (London: Verso, 1983). For a more recent view that pays special attention to the way constructs of national identity often exclude some citizens and stories, while amply including falsehoods and distortions, see Kwame Anthony Appiah, *The Lies That Bind: Rethinking Identity* (New York: Liveright, 2018).
12 David Burnham, *The Rise of the Computer State: The Threat to Our Freedoms, Our Ethics, and Our Democratic Process* (New York: Vintage, 1983); e-book, Open Road Distribution (2014), 48.
13 https://harvardharrispoll.com/wp-content/uploads/2023/05/HHP_May2023_KeyResults.pdf.
14 https://en.wikipedia.org/wiki/Telescreen.
15 George Orwell, *1984* (New York: Berkley, 2023), 3.
16 https://www.forbes.com/2009/08/10/government-internet-software-technology-breakthroughs-oreilly.html.
17 https://blog.pol.is/pol-is-in-taiwan-da7570d372b5.
18 https://blog.pol.is/uber-responds-to-vtaiwans-coherent-blended-volition-3e9b75102b9b.

INDEX

ABOUT THE AUTHORS

REID HOFFMAN is the cofounder of LinkedIn, cofounder of Inflection AI, and a partner at Greylock. He currently serves on the boards of companies such as Aurora, Coda, Entrepreneur First, Microsoft, and Nauto. He also serves on nonprofit boards, such as Kiva, Endeavor, CZI Biohub, New America, Opportunity@Work, and the MacArthur Foundation's Lever for Change. He is the co-host of the *Masters of Scale* and *Possible* podcasts. He is the coauthor of five bestselling books: *The Startup of You, The Alliance, Blitzscaling, Masters of Scale,* and *Impromptu.* He earned a master's degree in philosophy from Oxford University, where he was a Marshall Scholar, and a bachelor's degree with distinction in symbolic systems from Stanford University.

GREG BEATO has been writing about technology and culture since the early days of the World Wide Web. His work has appeared in *The New York Times, Wired, The Washington Post, The International Herald Tribune, Reason, Spin, Slate, BuzzFeed, The Guardian,* and more than 100 other publications worldwide.